Generic Construction of Availability Calculation Models for Safety Loops in Process Industry

Generische Erzeugung von Verfügbarkeits-Rechenmodellen
für Sicherheitskreise der Prozessindustrie

vom Fachbereich Elektrotechnik und Informationstechnik
der Technischen Universität Kaiserslautern
zur Verleihung des akademischen Grades

Doktor der Ingenieurwissenschaften (Dr.-Ing.)

genehmigte Dissertation

von

Thomas Gabriel
geb. in Bonn

D 386

Eingereicht am: 27.08.2010
Tag der mündlichen Prüfung: 30.11.2010
Dekan des Fachbereichs: Prof. Dipl.-Ing. Dr. Gerhard Fohler

Promotionskommission:
Vorsitzender: Prof. Dr.-Ing. Ralph Urbansky (Technische Universität Kaiserslautern)

Berichterstattende: Prof. Dr.-Ing. habil. Lothar Litz (Technische Universität Kaiserslautern)
Prof. Dr.-Ing. Frank Schiller (Technische Universität München)

Bibliografische Information der Deutschen Nationalbibliothek

Die Deutsche Nationalbibliothek verzeichnet diese Publikation in der Deutschen Nationalbibliografie; detaillierte bibliografische Daten sind im Internet über http://dnb.d-nb.de abrufbar.

ISBN 978-3-8325-2892-8

Logos Verlag Berlin GmbH
Comeniushof, Gubener Str. 47,
10243 Berlin
Tel.: +49 (0)30 42 85 10 90
Fax: +49 (0)30 42 85 10 92
INTERNET: http://www.logos-verlag.de

Abstract - Currently, safety instrumented systems (SISs) are mathematically treated by referring to pre-derived calculation formulas for SIS parts or by constructing suitable calculation models from scratch, using the recommended mathematical methods suggested by international safety standards. The former dramatically limits the degrees of freedom for the engineering process and requires significant simplifications of the SIS structure in order to make the formulas applicable. The latter quickly leads to large and complicated mathematical structures. Additionally, the standards do not provide information on how to implement a large variety of relevant aspects of modern SISs with these permitted mathematical approaches. As a solution to these problems, a generic approach is proposed in this work that automatically constructs suitable calculation models from a herefore invented formal description language, the Abstract Safety Markup Language (ASML). Two positive effects arise from that: firstly, all inherently and explicitly contained complexity of SISs is identified once and condensed into a set of transformation formulas generating the actual calculation models. Secondly, providing a flexible and extensive description language encourages smart and quick engineering processes, enables for non-standard solutions, and relieves safety engineers from the challenge of actually deriving explicit calculation models by hand. A suitable new type of discrete time multiphase markov model is chosen as mathematical basis, the Safety Multiphase Markov Model (SMPMM). It is introduced along with the required solver algorithms for retrieving the desired unavailability characteristics, including the average probability of failure on demand (PFD), as well as the probability of fail-safe (PFS), i.e., an operational unavailability as economical indicator.

Zusammenfassung - Die mathematische Behandlung Sicherheitsinstrumentierter Systeme (SIS) geschieht gegenwärtig durch Rückgriff auf verfügbare Standardformeln, oder aber durch händisches Aufstellen von Verfügbarkeits-Rechenmodellen auf Basis der durch die Sicherheitsnormen autorisierten Methoden. Ersteres schränkt die Freiheitsgrade beim Entwurfsprozess klar ein und erfordert signifikant vereinfachende Annahmen. Letzteres führt schnell zu großen und komplexen mathematischen Strukturen. Hinzu kommt, dass die Standards keine Hilfestellung bei der Modellierung einer Vielzahl relevanter Aspekte in modernen SIS bieten. Die vorliegende Arbeit bietet eine Lösung dieser Probleme durch die Einführung eines generischen Ansatzes, der auf Basis einer formalen Beschreibungssprache für SIS (Abstract Safety Markup Language - ASML) automatisch geeignete Rechenmodelle erzeugt. Zwei positive Aspekte erwachsen hieraus: Zum einen wird sämtliche inhärent und explizit in modernen SIS enthaltene Komplexität einmalig analysiert und in die zugehörigen Generatorformeln zum Erzeugen der Rechenmodelle überführt. Zum anderen ermöglicht eine flexible und umfangreiche Beschreibungssprache intelligente und schnelle Engineering-Prozesse, fördert kreative Lösungen, und befreit Sicherheits-Ingenieure von der Notwendigkeit, Rechenmodelle händisch erstellen zu müssen. Eine neue Art zeitdiskretes Markov-Modell dient als mathematische Basis, das Safety Multiphase Markov Model (SMPMM). Es wird zusammen mit den benötigten Solver-Algorithmen eingeführt, die die Ableitung der benötigten Unverfügbarkeitskenngrößen ermöglichen. Hierzu gehört die mittlere sicherheitstechnische Unverfügbarkeit (Probability of Failure on Demand - PFD), sowie die mittlere betriebstechnische Unverfügbarkeit (Probability of Fail-Safe - PFS) als öknomischer Indikator.

Acknowledgements

The work that resulted in this book could not have been accomplished without several persons' assistance, support, and encouragement.

First of all, I would like to thank my doctoral advisor, Professor LOTHAR LITZ, who offered me the opportunity to join his team, the Institute of Automatic Control at the University of Kaiserslautern. From the beginning of my work, he provided encouragement, guidance, as well as intellectual freedom. I am very grateful for his trust and advice that always turned out to be of great value for me.

Besides to Professor LITZ, my sincere thanks go to the members of the evaluation board: to Professor FRANK SCHILLER, and to the chairman Professor RALPH URBANSKY.

The time as a research assistant flew by in an instant, and I will always keep it in nostalgic memory. I thank all of my fellow colleagues for their cooperation and support, their motivation and productive discussions, as well as their friendship and honesty.

Over the years at the institute, numerous research projects with Bayer MaterialScience AG have been conducted under the administration of Dr BERND SCHROERS. I would like to warmly thank him for his memorable, open, and friendly advice, as well as the invaluable amount of inspiration that significantly contributed to this work.

Moreover, my thanks go to Dr DANIEL DÜPONT, Dr ANDREAS HILDEBRANDT, UDO HUG, KONSTANTIN MACHLEIDT, and Dr MATTHIAS ROTH for uncountable, very important discussions concerning my work.

To ALAIN CHAMAKEN, Dr MARTIN FLOECK, KONSTANTIN MACHLEIDT, STEFAN SCHNEIDER, and THOMAS STEFFEN for proofreading the manuscript and hinting at a considerable number of flaws. Extra credits go to Dr ROTH who fought his way through chapters 5.1 to 7.2 while retracing and challenging each and every formula.

Finally, I give the heartiest gratitude to my family: to my wife JACQUELINE for her love, her patience, and her encouragement. To my two girls TIBELYA and TABITHA for enriching my life with so much joy and pride.

Bergisch-Gladbach, May 2011

THOMAS GABRIEL

Contents

Chapter 1

Introduction

1.1 Motivation

After severe incidents in chemical plants such as Seveso (1976) or Bhopal (1984), safety management for the process industry sector has become more and more guided by international standards. Their intention is to provide a safety lifecycle for chemical plant sites that enables for plant operation at a sufficient level of safety. 'Sufficient level' equals a state with tolerably low risk of a severe incident. Over the last decades analyses as conducted by the U.S. Chemical Safety Board [Saf] have shown that human error in combination with poor maintenance and alarm management often initiate a chain of cause and effect which finally led to prominent incidents such as the Houston BP refinery explosion in 2005. Consequently, the quota of automated protection measures with rigidly organized maintenance plans increases constantly. These safety loops - their implementation is called safety instrumented system (SIS) - are usually completely detached from the distributed control system (DCS), providing as few manipulative interfaces as possible. This separation is implemented in the form of detached control cabinets, power supply, measuring and actuating devices, as well as independent maintenance teams etc. Each safety loop performs a protective function, i.e., its safety instrumented function (SIF), which is typically implemented in reliable standard technology such as 4-20mA current loops for measuring equipment and triggering of mostly pneumatic actuators. A typical SIS (see fig. 1.1 for an exemplary piping & instrumentation diagram) can therefore be subdivided into three parts. The sensor part is responsible for gaining information on relevant process variables. The logic solver part evaluates the single or multiple sensor signals from the sensor part and generates an output signal utilized for triggering the final elements. The final element part is intended to bring the process to a safe state by directly triggering appropriate actuators. Figure 1.2 depicts the device structure for the SIS from fig. 1.1, outlining the different system parts. Devices are typically arranged in channel structures, forwarding and/or transforming signals (see, e.g., the sensors, and the respective barriers in fig. 1.2).

A modern SIS is designed with the intention of maximizing two conflictive parameters: safety related and operational availability. The former is subject to rigid requirements,

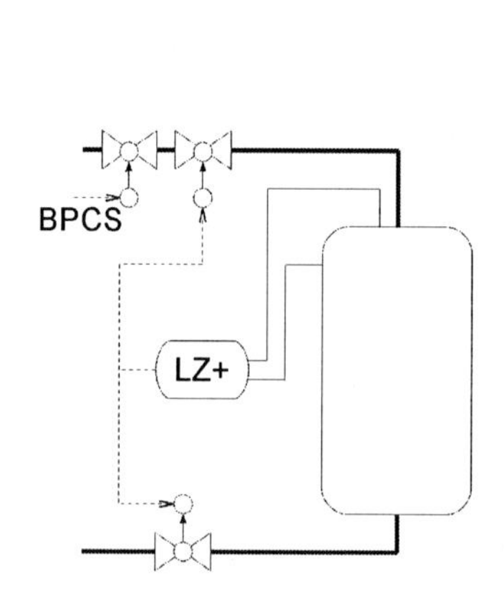

Figure 1.1: P & ID for an exemplary SIF with redundant sensors

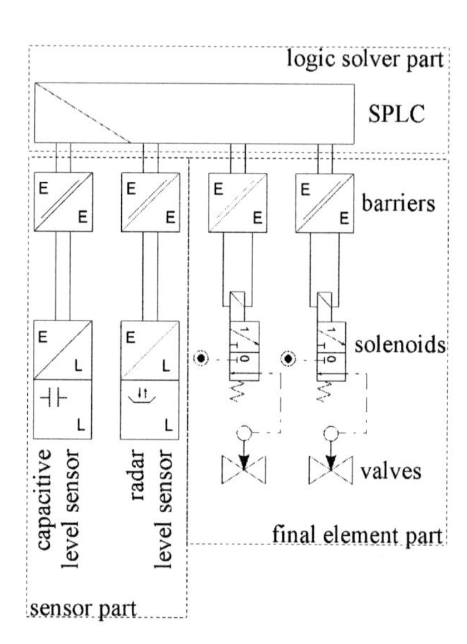

Figure 1.2: Device structure related to fig. 1.1

provided by various international safety standards, whereas the latter is subject to business plans and management decisions. Finding the optimal solution to this multi-variable optimization problem is difficult, since the numerical derivation of both parameters with regard to all relevant influencing effects and mutual interdependencies is difficult.

Numerical proof of the safety related availability is mandatory since IEC 61511 was released in 2003. Since those days it has become more and more clear that the standard's support for mathematically handling modern SISs is insufficient. Even simple safety loops require massive simplifications in order to allow for the utilization of the standards' calculation formulas. Due to the safety context, these simplifications tend to be conservatisms. As a result, being considered safe SISs require to be extended with expensive additional measuring or actuating equipment. This is critical especially for older sites that undergo a safety reassessment. Diverse or heterogeneous instrumentation, strongly encouraged by IEC 61511 [DIN05] in order to minimize the influence of common cause failures, may serve as an illustrative example for these unnecessary conservatisms. This concept can be implemented by, e.g., mounting two redundant measuring channels with different sensors and therefore different failure rates (compare the sensor part in fig. 1.2). It appears to be absurd that it is not possible to mathematically treat such diversely configured channels entirely. The user is forced to do the calculation with a substitutionary system which is homogeneously equipped with the failure rate of the less reliable channel. The effort put into the design of the original safety loop is completely wasted.

Besides this problem of unnecessary conservatism, additional severe problems remain unsolved. It is nowadays nearly impossible to mathematically handle safety loops that differ from the standardized architectural and behavioral patterns described in IEC 61511 and IEC 61508. There, the largest systems with respect to the number of channels are two-out-of-three (2oo3) structures. Nearly every burner shutdown SIS requires to actuate more than three valves which are usually arranged in redundant pairs due to the related high process risk.

Beyond the calculation formulas, the standards authorize the use of certain modeling techniques suitable for performing the required safety calculations, e.g., fault trees, and reliability block diagrams. These methods provide much more flexibility and allow for the integration of a large variety (depending on the considered method) of effects that are characteristic for modern SISs. However, these approaches are in almost all cases unfeasible for plant operators. The authorized methods require highly educated safety personnel with strong mathematical background. The models tend to become very large and confusing, even for smaller SISs. The manual modeling process is exhausting and prone to failures. Simple modeling techniques are not capable of reproducing important effects such as, e.g., dynamic variations in SPLC (safety programmable logic controller) voting algorithms. Advanced techniques tend to result in complex and abstruse models very quickly. No instructions on how to precisely model certain effects with the provided techniques are given. From this, a strong legal uncertainty arises, preventing plant operators from utilizing these mathematical tools.

Since the standards' primary intention is to ensure sufficient plant safety, the occurrence of false trips is implicitly accepted and thus not evaluated. This fact totally contrasts the claims of plant operators which is - as mentioned in the beginning of this section - the optimization of both safety related and operational availability.

1.2 Aim of the thesis

This thesis aims at solving the previously described problems by introducing a new approach for the generic construction of availability calculation models for safety loops in process industry (see fig. 1.3 for an overview).

In a first step, the safety engineer provides a formal specification of architecture, instrumentation and behavior of the SIS. This is accomplished by utilizing the newly introduced 'abstract safety markup language' - ASML. ASML provides a mostly graphical workflow without need for any sort of high-level programming language, thus rendering it a flexible tool with easy access for engineers. This first step is also the last one requiring any engineering effort. The mathematical object indirectly specified by an ASML description is called an ASMLSIS.

The second step of the ASML approach is the automated synthesis of calculation models. This step depends on a strict and complete parameterization of the provided transformation formulas which is assured by an ASML description. A newly introduced type of discrete time markov model serves as mathematical background for the calculation model: the 'safety multi phase markov model' (SMPMM), which is identified as the most

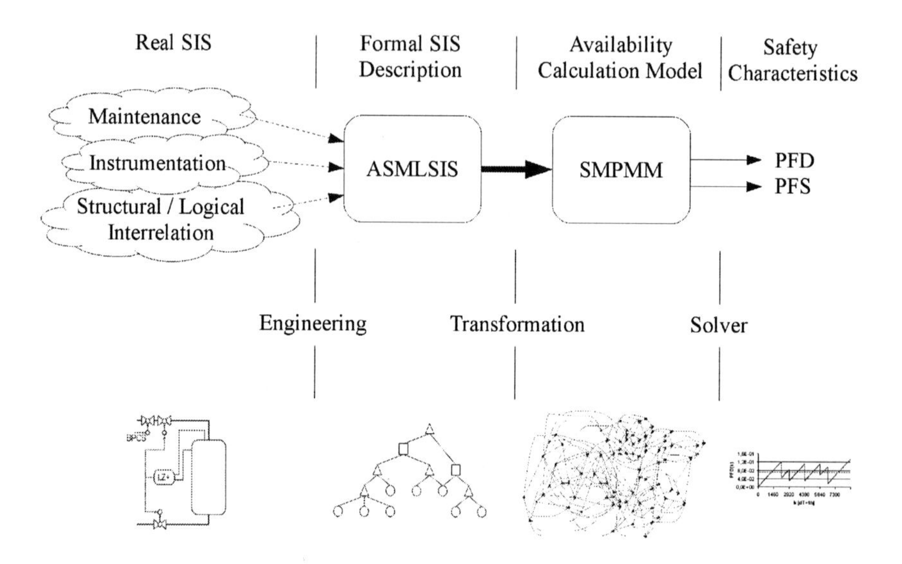

Figure 1.3: Overview of the ASML approach

suitable modeling technique with regard to the requirements arising from this work's motivation.

From such an SMPMM, the desired availability characteristics can be obtained in the third and last step. In addition to the standardized safety related PFD (probability of failure on demand), an operational unavailability characteristic is introduced: the PFS (probability of fail-safe). It refers to a spurious trip state of the overall SIS, i.e., having the system perform its predefined safety task unintendedly. The PFS contributes to satisfying the operators' demand for economical SIS optimization, as will be explained in the course of this work.

The two main steps of this thesis - providing a formal description language for SISs as well as suitable transformation formulas - are of equal importance. Both aspects together result in a powerful framework - called the ASML approach - in the shape of a toolchain enabling for flexible and easy calculation of arbitrary modern SISs under consideration of state-of-the-art technological effects.

The ASML approach sets itself apart from the few available comparable approaches. As ASML is utilized to describe the SIS rather than the explicit calculation model, it differs, e.g., from [But86], and [JB88]. There, a formal description language allows for the compressed specification of an explicit markov model's state space, transitions etc. . It allows for a short notation of the features of a special class of markov processes, but requires full engineering effort in order to set up the target model first. [PR99] on the other hand provides a SIS description on the basis of constrained automata. This powerful framework enables for advanced synchronization effects among specified components as well as for flexible behavioral patterns. But as the system behavior is described using some kind of high-level programming language, the automated transformation into calculation is difficult. Under several simplifications it is possible to generically derive basic (static) fault tree models. But as these models lack relevant required capabilities, and as the approach demands extended expert knowledge, it is not applicable within the scope of this work. [GY08] does not introduce an explicit description language. Instead, the user of the related approach directly parameterizes the provided transformation formulas which generate markov models of the commonly applied type in the safety community (multiphase markov models). The authors adopt most of the assumptions from the relevant safety standards. They therefore assume, e.g., a sensor channel, to be 'atomic', i.e., without internal structure. An additional significant restriction in [GY08] are homogeneously instrumented subsystems, i.e., having all sensor channels involved in a voting be identical in structure and behavior, which is unrealistic for real SISs. The subsequent sec. 2.2 (p. 10) will introduce several interesting effects occurring in modern SIS that require an engineering process on the basis of individual components rather than channels (which are covered by the ASML approach) as well as heterogeneous instrumentation.

1.3 Structure of the thesis

The subsequent chapter 2 provides an overview of concepts, strategies, structures and relevant effects in modern SISs (sec. 2.2 and sec. 2.3). The herefrom gathered information is condensed into a list of requirements to the formal description language (sec. 2.4).

Chapter 3 provides the basic mathematical access to unavailability calculation. Starting from the normative context (sec. 3.1 and sec. 3.2), the central terms 'unreliability' (sec. 3.3), 'maintenance' (sec. 3.4), and 'unavailability' (sec. 3.5) are derived. The most important unavailability cycles, consisting of failure and repair phases, are explained and put into relation to the identified challenges from chapter 2. The two target safety characteristics PFD (probability of failure on demand) and PFS (probability of fail-safe) are explained or defined, respectively. This chapter is the fundament for all subsequent declarations as important terms and operations are introduced.

Based on the general unavailability principles, the most important authorized (i.e., explicitly encouraged by the relevant safety standards) mathematical calculation approaches are introduced in chapter 4 (sec. 4.2 to sec. 4.5). Referring back to the list of requirements from 2, a critical evaluation of the approaches' suitability for the generic ASML approach is conducted (sec. 4.6). The chapter closes with the well-founded choice of markov models as the favored calculation approach.

Chapter 5 introduces the formal SIF description language ASML (abstract safety markup language). As mentioned above, the central object is the ASMLSIS (sec. 5.4) which has the ASML graph ASMLG as its most important integral part (sec. 5.3). The basic thoughts and concepts behind the ASML approach are recapitulated in sec. 5.2. Here, most of the basic mathematical declarations are made.

The set of generator equations - performing the synthesis of calculation models based on an ASMLSIS - can be found in chapter 6. It can be subdivided into the declaration of a new type of markov model, the SMPMM (sec. 6.2), as well as the generator formulas (sec. 6.3). Along with the SMPMM come the required formulas for retrieving the desired safety characteristics PFD and PFS from the markov model.

As the ASML approach is intended to be applicable for real SISs, a software implementation is conducted. Along with the implementation, a strong need for program optimization methods arises (chapter 7). As markov models grow exponentially with SIS size, suitable methods need to be provided that keep the resulting models executable on standard PCs. Most important approach is the transformation of an ASMLSIS into a DASMLSIS, i.e., a decomposed ASMLSIS. This method makes use of stochastically independent SIS parts (sec. 7.2) and allows for a significant but lossless reduction of model size and thus memory consumption as well as computational effort. Additional optimization methods, such as the lossless state space reduction (sec. 7.3) or the sparse matrix implementation (sec. 7.4) allow for further reduction of computation time and memory consumption.

It is certainly necessary to provide a sufficient amount of verification and validation of the proposed methods. As it is impossible to validate the generated calculation models with real test scenarios (which would require an unrealistically long observation period), several validation methods are chosen in chapter 8. These include cross checks with qualified sources (sec. 8.3), a priori knowledge validation (sec. 8.2) and the evaluation of ASML-specific effects (sec. 8.5).

The thesis is concluded in chapter 9 with a summary and an outlook.

1.4 Remarks

This thesis refers to spurious trips of components or subsystems as 'active' failures rather than 'safe' failures. This derives from the definition of dangerous failures in [DIN02]. There, all failures that are not 'dangerous' are automatically classified as 'safe'. Safe failures therefore refer to 'no effect' failures, 'annunciation failed' as well as 'active' failures (see, e.g., the failure classification in [exi]). In this total safe failure fraction, only the active failures are relevant for this work.

"Definitions" are used to indicate new functions, algorithms, sets, characteristics or parameters that have not yet been defined in other sources or had to be reparameterized fundamentally for this work. Variations of material from other authors refer to the original work by citation. Slight extensions to common knowledge are not specifically highlighted.

Elements of vectors are denoted as $\boldsymbol{a}[i]$, representing the i-th element of vector $\boldsymbol{a}$.

All stochastic processes mentioned in this thesis are lifetime processes and therefore not defined for $t < 0$. Thus, the piecewise definition of distributions and densities has been shrunk to the definition of the part $t \geq 0$. All exceptions are outlined.

The number of elements in an arbitrary vector is denoted with $|\boldsymbol{v}|$.

All vectors used throughout this thesis are introduced as column vectors. If a row vector is required, it is denoted as $\boldsymbol{c}^{\top}$. The same notation is used if a vector is to be denoted explicitly. An exemplary column vector $\boldsymbol{a}$ reads as $\boldsymbol{a} = (a_1\, a_2\, a_3)^{\top}$.

The evaluation result of a function $\mathrm{f}(x)$ mapping $(\cdot) \mapsto \{true, false\}$, can be abbreviated according to, e.g.,

$$y = \begin{cases} 1, \text{ if } \mathrm{f}(x) \\ 0, \text{ if } !\,\mathrm{f}(x) \end{cases} ,$$

instead of

$$y = \begin{cases} 1, \text{ if } \mathrm{f}(x) = true \\ 0, \text{ if } \mathrm{f}(x) = false \end{cases} .$$

Throughout this work the internal states such as OK or DU (dangerous undetected) of arbitrary components are denoted in capital letters. Lowercase is used for the related failure types such as dd (dangerous detected) or a (active).

The represented stochastical failure and repair processes utilized in this thesis are continuous time processes. The applied markov model techniques use discrete time models. These are intended to approximate the real processes as well as possible. In all figures depicting markov characteristics over discrete time, the curves are drawn continuously in order to refer to the fact that the underlying process is not discrete. The step size Δt is chosen small enough to ensure a sufficient model accuracy.

Chapter 2

Challenges and characteristics in modern SIF design

2.1 Overview

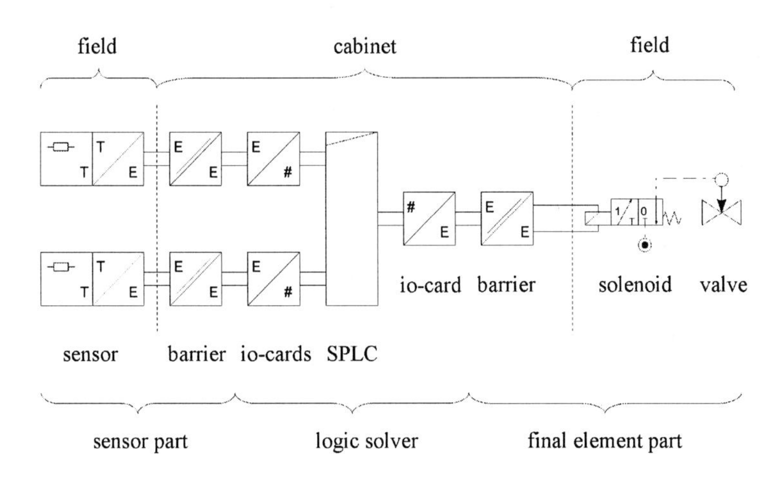

Figure 2.1: Structure of a simple SIF with redundant sensor part

Up to nowadays the basic tasks of SISs have remained the same. By utilizing suitable measuring devices a dangerous process state is to be detected. In case of such a 'demand', the SIS is responsible for bringing the process to a safe state which is accomplished by actuating certain final elements. These final elements then shut feed pipes, segment pipe sections from each other, release pressure from vessels etc.

The basic structure of a SIS can be subdivided into a sensor part, a logic solver part, and a final element part as defined in [DIN05]. The example in fig. 2.1 shows a SIS implementing a temperature protection. It consists of a redundant temperature measurement, using two PT100s connected to transmitters. The 4-20mA signal connects to the logic solver via barriers and io-cards. The final element part is not redundant and consists of io-card, barrier, solenoid and a valve as final element. Despite the subdivision into said system parts, a categorization can also be conducted by referring to the location of the various devices. While some components are field devices (sensors, solenoid, valve), others are mounted in cabinets.

It is important to stress on the fact that the safety related quality of a SIS not only depends on the failure behavior of the underlying components. Maintenance has a major impact. Typically, a chemical plant is scheduled to be shut down approximately once per year. This stop has the intention of conducting preventive as well as corrective maintenance for the process equipment such as replacing filters, cleaning nozzles or burners. To simplify matters, the process safety department uses this time window to perform proof tests with the safety equipment. The aim is the revelation of hidden failures in the shutdown path that would otherwise prevent the SIS from performing its intended safety function. These proof tests during plant shutdowns are very effective: as the whole safety function can be tested by initiating a demand at the sensor part and monitoring the final element part's reaction, a large fraction of hidden failures can be found.

Most recent trends aim on extending and generally adapting the proof test intervals by, e.g., lowering the test frequency for very reliable components. This and further specific maintenance strategies for all three SIS parts will be outlined subsequently. They serve as basic requirements to the ASML approach which is capable of reproducing such concepts.

2.2 Structure of modern SISs

2.2.1 Sensor part

Standard current signal

Communication among components of the sensor part of a SIS (and large fraction of the final element part) is based on the 4-20mA standard current signal as introduced in, e.g., [Win08] (3. 2). The valid range for encoding measurement values is a current value between 4mA as lower limit, and 20mA as upper limit. A signal below or above these limits is interpreted as an erroneous signal. In case of, e.g., an open circuit failure the current drops down to 0mA, clearly indicating the failure. This 'life-zero' principle additionally allows for continuous energy supply of the involved components, as a current >0mA is always required for valid signal values. Accordingly, for short-circuit failures, a current beyond 20mA results. The standard current signal has been used in the chemical process industry for a long time, as it provides ideal characteristics for safety applications: high reliability, simplicity, maintainability, scalability, and applicability under various environmental conditions. Due to these positive characteristics, more sophisticated communication methods such as field buses are usually not in service for SISs.

Instrumentation and structure

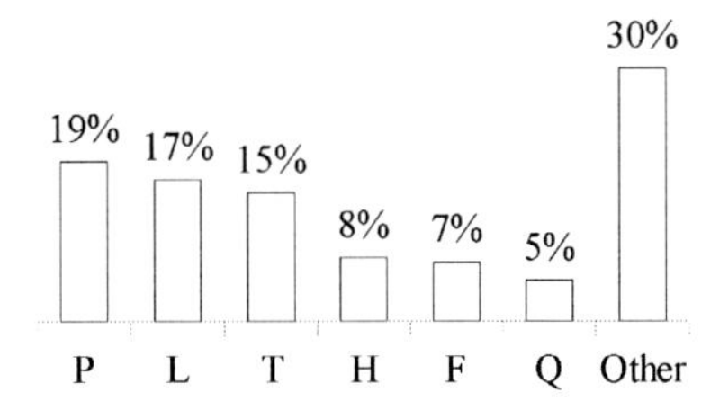

Figure 2.2: NAMUR measuring principle distribution 2008 in single channel SISs according to [D"u10]

Sensors in safety relevant applications of the process industry can be categorized by their respective monitored product parameter. According to recent NAMUR investigations, summarized in [D"u10], the largest fraction of measuring devices is installed for temperature (T), pressure (P), flow (F), and level (L) monitoring (see fig. 2.2). Around 60 percent of all sensors in single channel SISs origin from these four categories. The remaining 40 percent consist of a larger fraction of applications with manual operator interaction (H), quality measurement (Q), and others. The latter ones refer to, e.g., proximity measurement or flame monitoring. Quality measurement refers to the analysis of chemical product parameters such as e.g density, conductivity, oxygen content, opacity or pH value. Such systems are characterized by a high degree of complexity. Process analysis technology (PAT) sensor installations consist of measuring devices supported by sample drawing, analysis and disposal, calibration procedures, subordinated redundancies and specialized mini SPLCs for measurement control. Most sensors are selectively available with analog or binary output signal, e.g., level or proximity switches.

Sensors can be subdivided into a sensor element and a signal converter that transforms the individual raw signal into the standard 4-20mA current signal. Several components such as feed units or different types of transmitters like isolation transmitters provide energy supply, explosive protection, galvanic isolation, and - if required - additional signal conversion (see figs. 2.3 and 2.4). These components are usually arranged in channel structures, i.e., sequential device chains. Thus, the information about the desired process parameter is forwarded to a subsequently allocated SPLC where it is evaluated.

Relevant failure behavior

Sensor channel components are typically electrical/electronical/programmable electronic devices according to [DIN02], as they have to generate, transmit, or convert a current signal. The most relevant types of failure can be subdivided into four different categories, where the first three are described, e.g., in [exi] and the latter one has been explained in detail in [Wei09].

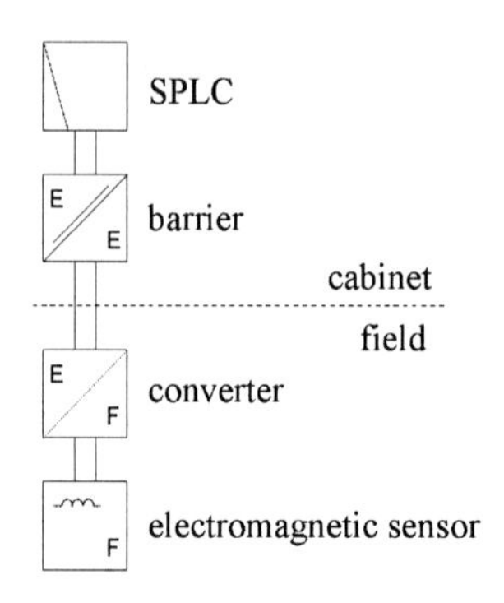

Figure 2.3: Simple flow sensor part

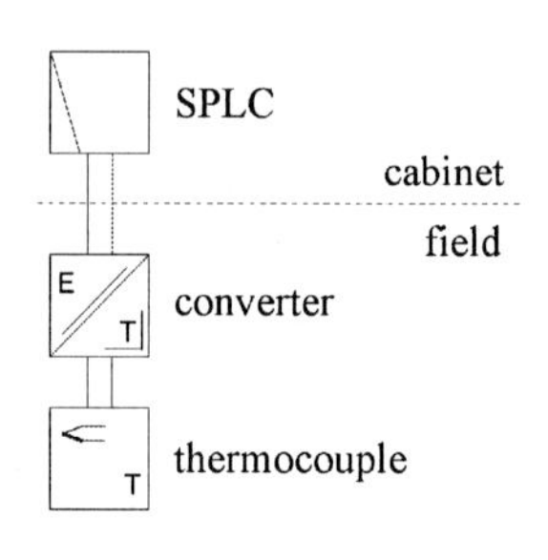

Figure 2.4: Simple temperature sensor part

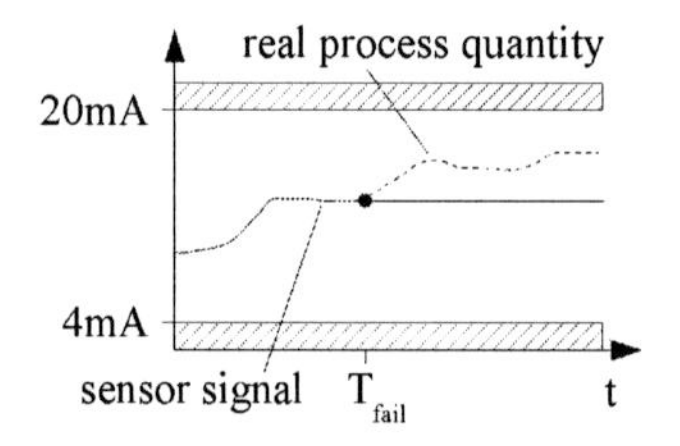

Figure 2.5: Output of sensor component with dangerous undetected failure

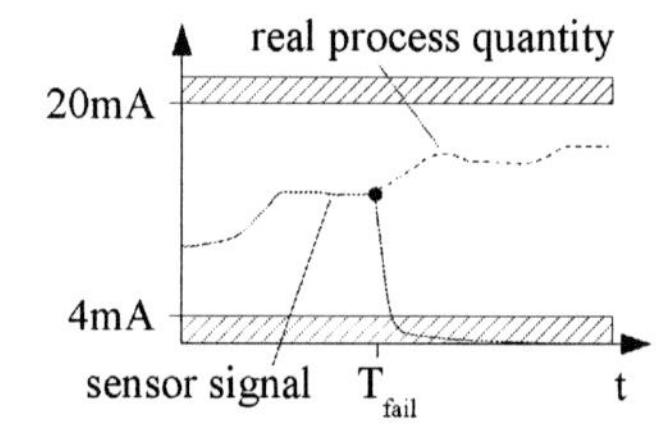

Figure 2.6: Output of sensor component with underrange failure

- Dangerous undetected (*du*) failures prevent the component from forwarding a valid signal. The output current is stuck or erroneous in the signal range between 4 and 20mA (see fig. 2.5). As the output does not enter over- or underrange, a detection of this type of failure is impossible on the basis of the single current signal alone. A possible source for a dangerous undetected failure are damaged (stuck) CPU register bits mapped to the sensor output.

- Over- and underrange failures force the device's output to produce a signal value beyond one of the two failure ranges (<4mA or >20mA). In almost all cases the output drops or increases with a very high slew rate. The dynamics of the monitored process can typically not reproduce such behavior. This renders over- and underrange failures potentially detectable. A typical source for an underrange failure is, e.g., a failure in the power supply module, causing the output to drop to 0mA (see fig. 2.6).

- Most modern sensors are equipped with extended internal diagnostics. By performing, e.g., memory or plausibility checks, a large fraction of otherwise undetectable failures gets revealed. If the internal diagnostic circuits identify a failure, they force the output to either over- or underrange. The particularly chosen failure range is usually configurable and depends on the specific type or location of the detected failure within the device. With regard to the current signal only, failures detected by internal diagnostics can not be differentiated from over-/underrange failures.

- Additional diagnostic effort is put into certain applications. Secondary sensor devices without actual safety functionality monitor specific characteristics of the primary safety sensors. The gathered information is used for detecting failures that would remain unrevealed otherwise. The diagnostic information is typically not forwarded by the safety-relevant 4-20mA current signal but by additional communication channels such as field buses of the DCS. A complex oxygen PAT sensor may serve as an example for that. The primary oxygen sensor might get plugged, resulting in an erroneous signal output. An additional flow sensor is used to measure the flow through said device and thus provides an additional diagnostic measure for the specified type of failure. Additional sensor devices are also additional sources of failure which could be mathematically treated in the calculation models.

It is important to point out that the described failure types' actual impact on the SIS cannot be determined without considering the technical capabilities of the related SPLC as well as the safety task specification. A potentially detectable failure could be a *dd* or an *a* failure, depending on the context. Subsection 2.2.2 provides the additional considerations required for a suitable classification of the sensor device failures.

The sensor element is the only part of the whole measurement chain being in contact with process medium. Systematic failures such as wrong dimensioning or wearout effects may have great impact on its failure behavior. These sources of failure must therefore be prevented or at least under control. According to [DIN05], this topic is to be excluded from calculations and therefore not covered in the course of this work.

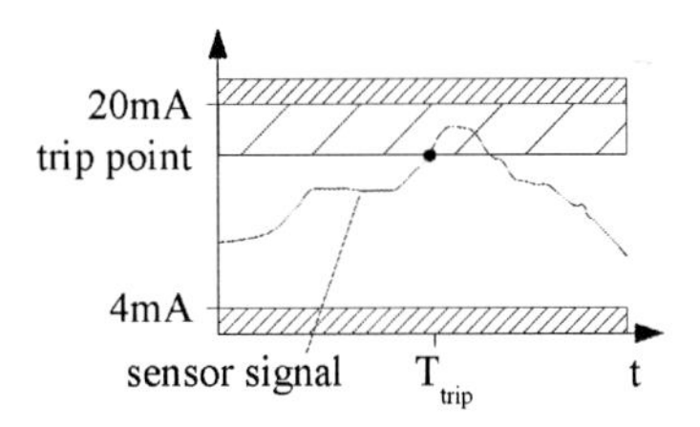

Figure 2.7: Sensor signal of a high trip SIS denoting a trip at T_{trip}

2.2.2 Logic solver

Basic principle

Safety programmable logic controllers (SPLC) or logic solvers (LS) are multi-redundant CPU modules with IO cards attached. Similar to standard PLCs in, e.g., manufacturing processes, they are intended to perform a periodically executed loop, the SPLC cycle. It consists of reading binary and analog input signals, evaluating them and setting outputs according to the desired control strategy. The evaluation algorithms on SPLCs range from simple one-input-triggers-one-output structures to complex programs with integrated process models. In most safety applications a voting algorithm is used in order to provide redundancy. The term 'MooN' as defined in [DIN02] refers - among others - to SPLC voting algorithms. In these algorithms, the N sensor input signals are continuously scanned for at least M trip signals in order to trigger the predefined final elements. This is implemented by defining sensor trip points, i.e., input signal values indicating the transition from a process quantity in tolerable range to its dangerous range.

Example 2.1. A SIS is implemented as an overheating protection for a reactor. The dangerous state is reached if the temperature goes beyond 150°C (T_{trip} in fig. 2.7). Thus, the two redundant sensors might be configured such that 4mA $\widehat{=}$ 10°C and 20mA $\widehat{=}$ 210°C. The related SPLC monitors both temperature sensor signals and evaluates whether they go beyond 16mA. This current value represents the specified trip point of 150°C, rendering the protection task a 'high trip', as the dangerous state is reached if a certain process parameter is higher than the specified trip point. The SPLC triggers its connected final elements if at least one of the two sensor channels trips, rendering the chosen algorithm a 1oo2 voting.

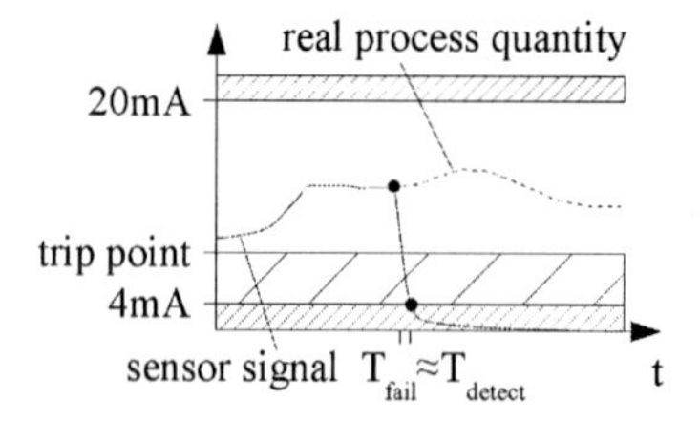

Figure 2.8: Low trip SIS sensor failing underrange

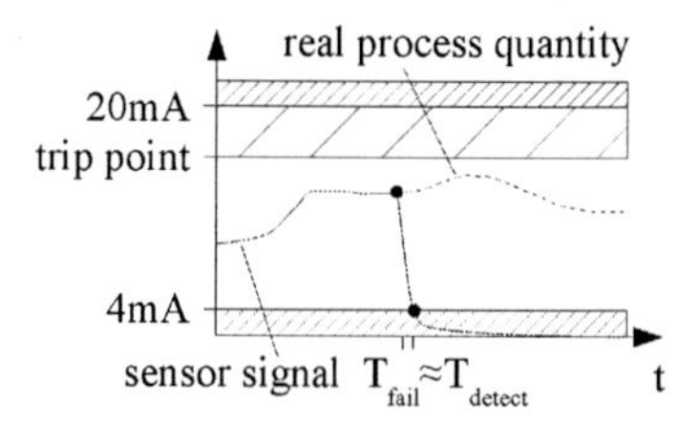

Figure 2.9: High trip SIS sensor failing underrange

Classification of sensor device failures

It is important to stress on the fact that the sensors are not 'aware' on whether they are connected to a low or a high trip SIF. This leads to the problem that the various failure modes of devices in sensor channels have different consequences depending on the implemented algorithm on the related SPLC. The resulting behavior depends on three important parameters:

- Transient analysis capabilities of the SPLC, i.e., identification of strong in- or decrease of the input signals which can only be caused by device failures rather than the underlying process
- Over- / underrange detection capabilities of the SPLC
- High or low trip configuration of the implemented SIF

Consider a situation according to fig. 2.8. The only sensor of a low trip SIS fails underrange at T_{fail}. Its current drops to 0mA within very short time. Several situations are possible:

- If the SPLC is neither capable of over- / underrange detection nor transient analysis, the sensor failure results in a spurious trip of the SIS, as the sensor channel signal goes beyond the defined trip limit. Since over- / underrange detection is not available, the origin of the spurious trip is unknown. It is likely that plant operators will expect a valid shutdown due to a demand at first. The responsible underrange failure of the sensor channel is therefore to be classified as 'active undetected' (au).
- If the SPLC is capable of over- / underrange detection but not transient analysis, the sensor failure again results in a spurious trip of the SIS. But as opposed to the first case, it is detectable now at T_{detect}. The SPLC may announce that a sensor channel failure led to the plant shutdown, resulting in short repair time. The failure is classified as 'active detected' (ad).
- If the SPLC is capable of over- / underrange detection and transient analysis, the sensor failure is detected by the transient analyzer *before* the signal goes beyond the trip point. It is therefore possible for the SPLC to not trigger the safety function when the signal exceeds the trip point and only announce the sensor failure. The sensor channel behavior has to be classified as 'dangerous detected' (dd).

Another important scenario is given in fig. 2.9. Here, the sensor channel fails underrange again, while the SIS is configured as high trip. As long as underrange detection is available at SPLC level, the occurring failure is called dangerous detected (dd) as long as the failure is announced only: the sensor channel does not provide valid measurement, but as the trip point is not exceeded no false trip is initiated. If the logic solver is programmed in a way that it triggers the final elements in case of a detected failure, the sensor's underrage failure must be classified as active detected (ad). In case of missing underrange detection on the other hand, the failure would be called dangerous undetected (du).

More complex situations can easily be constructed if the SPLC is, e.g., capable of either overrange or underrange detection only. In case of dangerous undetected failures (see fig. 2.10) no configuration variants exist. The failure is undetectable at SPLC level and therefore remains dangerous undetected (du).

A general overview on how sensor failures have to be interpreted on SPLC level is shown in table 2.1, following [exi]. The underlying assumption is that range detection

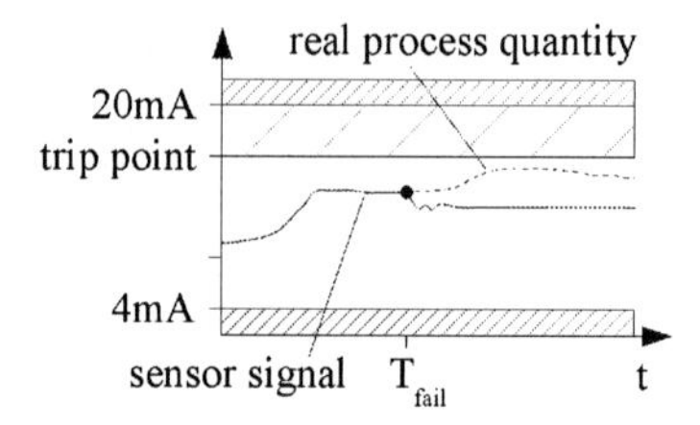

Figure 2.10: High trip SIS sensor failing dangerous undetected

Table 2.1: Interpretation of sensor failures at SPLC level following [exi]

Range detection	Transient analyzer	Trip point	Fail low	Fail high
Yes	Yes	High	dd	dd
Yes	No	High	dd	ad
Yes	Yes	Low	dd	dd
Yes	No	Low	ad	dd
No	No	High	du	au
No	No	Low	au	du

includes both over- and underrange detection. Sensor channel component failures are only assumed to result in either the one or the other out of range failure. Additionally, it is assumed that if transient analysis is enabled, it is utilized to prevent the channel from tripping if the signal goes beyond the trip point.

Although active detected and active undetected failures may formally be differentiated from each other, their consequences with respect to the SIS behavior are very similar. Both initiate at least a sensor channel trip. Depending on the implemented voting, both have the potential of bringing the plant to a safe state. In that case the time to identify the failure origin is negligible compared to the actual time for the startup process required after a shutdown. In case of redundant sensor channels with voting algorithm using $M > 1$, a single channel failing actively will not result in an immediate SIS trip. Even if the SPLC has no over- / underrange detection capabilities: the single active failure is detectable by comparing all signal values with each other. A single deviating signal is easily identified and repaired quickly. Apart from that, almost all modern SPLC provide over-/underrange detection as well as transient analysis. It is therefore legitimate to treat all active failures detectable as long as the related over-/underrange signal is forwarded correctly to the SPLC (which depends on the failure state of subsequently allocated components as shown below).

It is important to stress on the fact that single active sensor device failures alone do not necessarily dictate the overall state of the SIS. If multiple channels are available and the SPLC runs a voting algorithm with $M > 1$, a single active failure is not sufficient to get the SPLC outputs triggered. For this reason, 2oo3 votings are preferred over 1oo2 if operational availability is important. They have very similar safety related availabilities, but in 2oo3 systems at least two sensor channels need to fail actively before a SIS false trip is initiated (which is very unlikely, as most active failures are detected and repaired within short time).

Relevant failure behavior

The fraction of dangerous undetected (*du*) failures for SPLCs is typically very low due to extended internal diagnostics based on, e.g., redundant CPUs with cross check methods (see [B"o04]: 5–8). This leads to the good engineering practice of neglecting logic solvers in PFD calculations, as their safety related unavailability is normally at least one order of magnitude smaller than the rest of the equipment. Most dangerous failures are detectable with configurable consequences. If failures are announced only, they have to be considered dangerous detected (*dd*). If they immediately initiate a final element trip, they behave like active detected failures (*ad*). Additional failure types are active undetected failures (*au*), causing spurious trips in the final element part while not being announced, e.g., failures in the output drivers.

Dynamic voting

The evaluation algorithms performed on the CPUs tend to grow more and more complex. Aside from the standard static voting algorithms as specified by IEC 61508 (e.g., 1oo2, 2oo3), modern dynamic voting is used. An important variation of dynamic voting is applied on a regular basis: degraded mode systems use different algorithms depending on the number of pending detected failures at the inputs. This method is typically used to account for the knowledge of having parts of the SIS temporarily not available. It does not make sense for a 2oo2-system to wait for two trip signals while one channel is knowingly dangerous detected. In such cases a valid option is to switch over to a 1oo1-voting, ignoring the unavailable channel. The safety functionality is kept while a repair is executed for the failed channel.

2.2.3 Final element part

Instrumentation and structure

Classic final elements for the process industry are pneumatic actuators in combination with cone valves (control valves), ball valves (see fig. 2.11) or butterfly valves. The related actuators are usually triggered via solenoids with current or voltage input, often in combination with additional solenoid control modules, forming channel structures just like sensor components. More and more electromechanical components such as contactors, motor starters, and relays (see fig. 2.12) are used as final elements. In rare cases an electrical pump or drive is configured as energize-to-trip, i.e., the safe state is not powerless state as normally preferred. In that case these complex devices become part of the SIS and require to be included into qualitative and quantitative considerations.

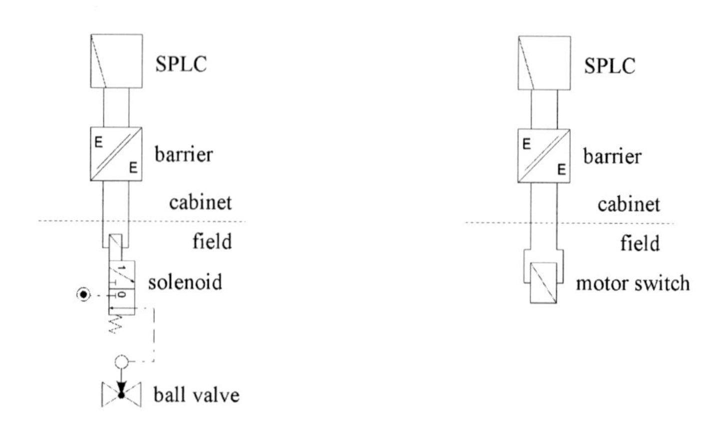

Figure 2.11: Final element part with ball valve

Figure 2.12: Final element part with motor switch

Relevant failure behavior

Mechanical as well as electromechanical components behave differently than electrical and electronic devices since their failure mechanisms are mainly based on wearout effects. The mathematical lifetime model assumed in [DIN02] serving as the basis for quantitative SIL assessment is therefore not appropriate. Despite these shortcomings, [DIN05], released in 2003, demands the quantitative assessment for complete safety loops, including final elements. Suitable calculation methods enabling for the correct modeling of said wearout effects are not provided. This problem is currently solved by using heavy (conservative) approximations that make mechanical components treatable with mathematical models for electronic devices. These approaches are presented, e.g., in [DIN07] and [SN 07].

The largest fraction of dangerous failures in final elements are dangerous undetected (du), as typically no diagnostic methods are available. Sometimes position switches are used to indicate whether an actuator or valve has failed to move to the desired position. But these devices do not render the du failure dd, as the detection is available in the moment of the demand at the earliest. Appropriate response is typically not possible.

Active failures are typically detectable, as they impact on the process which is monitored by the basic process control system (BPCS). A solenoid, failing safely, would deaerate the related actuator and thus bring the valve to its safe position. This would result in a notable effect on the process and finally lead to failure detection.

Partial valve stroke test

A very recent technology is the so-called partial valve stroke test, where an additional actuating device is used to minimally move the assigned final element on a regular basis

[MS06]. The related friction curve is analyzed and provides information on the status of the actuator-valve combination. Most implementations conduct this analysis in the BPCS, few locally by utilizing intelligent field devices with 4-20mA line to the SPLC. Mathematical treatment has been conducted, e.g., in [BSH08] and [LR08a]. A specific fraction of *du* failures is detectable by such partial proof tests. These failures will be referred to as *dup* (dangerous undetected partial proof test detectable) failures from now on.

2.3 General SIS effects

2.3.1 Diversity

If multiple sensor or final element channels are implemented, homogeneous as well as diverse instrumentation can be found. Homogeneous instrumentation refers to the utilization of identically constructed components. The degree of diversity can be adjusted by utilizing components from different manufacturers, mounting components in different locations of the process, using different measuring principles or even different physical quantities. The higher the degree of diversity, the lower the potential for common cause failures (see subsec. 2.3.6). Diverse instrumentation is encouraged by the safety standards, whereas the authorized calculation formulas only work for homogeneous structures. A strong need for suitable mathematical support arises from that.

2.3.2 Inter-channel connections

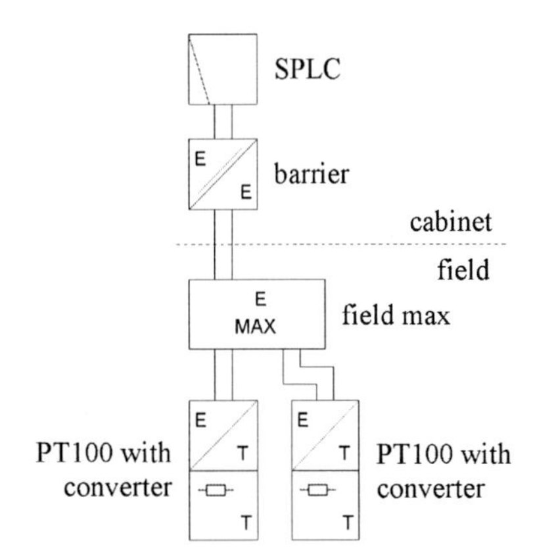

Figure 2.13: Field max example

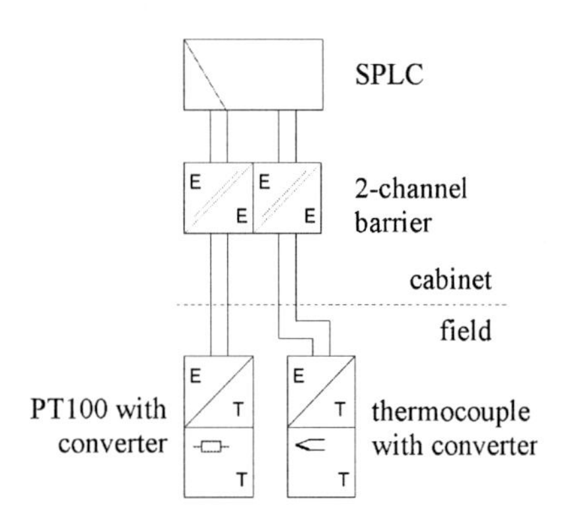

Figure 2.14: Multi channel barrier example

Inter-channel connections are possible, e.g., in the form of multi-channel barriers (see fig. 2.14), feeding several channels with only one device or field-max devices (see fig. 2.13) that perform a max choice from two current signals. On final element side, inter-channel connections refer to, e.g., a single solenoid driver feeding two solenoids without actually using a signal splitter or a single auxiliary relay triggering several main contactors. Neither IEC 61511 nor IEC 61508 consider this degree of freedom for SIS design explicitly. Nevertheless, systematic mathematical treatment is desirable, as such implementations are the most suitable solution to certain safety tasks.

2.3.3 Inhibition

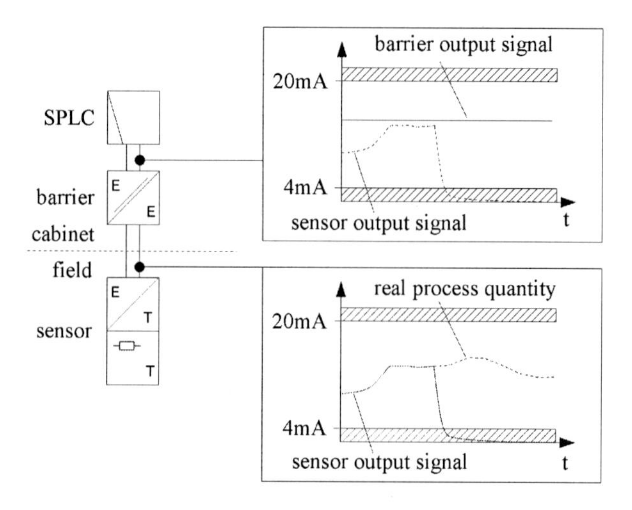

Figure 2.15: Active sensor device failure not reaching the SPLC: inhibition effect

An effect currently not considered by neither the safety standards nor research groups is 'inhibition'. An over-/underrange signal from a sensor component might not reach the related SPLC if a subsequently allocated component suffers from a *du*-failure. In fig. 2.15, the barrier component has failed *du* at an arbitrary point in time. Afterwards, the sensor element failed with a potentially detectable failure by going underrange. Since the barrier remains with constant output - regardless of its input decreasing to underrange - the SPLC only 'sees' a valid signal in the 4-20mA range. The otherwise *dd*-labeled failure of the sensor element is 'inhibited'. This effect is only observable if dependencies among individual components of a sensor channel are considered. The standards as well as the associated research activities focus on complete channels rather than individual components and therefore neglect inhibition. The impact of inhibition on the SIS's unavailability is denoted in the course of this work.

Obviously, inhibition effects may also occur in the final element part of a SIS. A solenoid failing actively cannot cause a spurious trip of the related final element if the subsequently allocated valve actuator has failed *du*, e.g., by getting stuck.

2.3.4 Secondary sensor implementation

Secondary sensors as introduced in subsec. 2.2.1 (p. 10) may be utilized as extended diagnostic devices for safety relevant primary sensors. The secondary sensor diagnostic information completely bypasses the SIS channel communication structure, as it uses its own 4-20mA line or different technology. The aforementioned inhibition effect is therefore not relevant for these. If secondary sensors are in their functional state, they render a specific fraction of sensor *du*-failures detected or active (if a detected failure initiates a channel trip). The standards and associated publications do not explicitly prescribe how to implement such secondary sensors. Sometimes the related communication is wired to the SPLC, sometimes to the BPCS (maybe controlling an annunciation in the operator station). It is not clarified under which conditions and how their functionality and failure behavior is to be treated in safety calculations. This work will provide a proposal for a solution to this problem.

2.3.5 Combinatorial voting

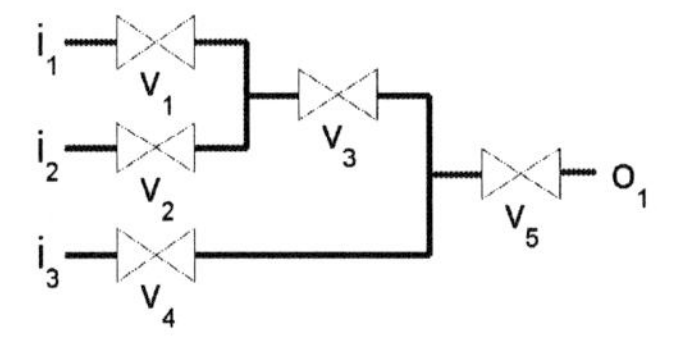

Figure 2.16: Example for non-MooN-redundancy

Combinatorial voting refers to systems where an MooN voting scheme is not sufficient to describe the applied redundancy. Consider the pipe header as shown in fig. 2.16. Such systems cannot be described with the MooN standard: The safety function requires the interception of medium flow from the left side to the right. Multiple combinations of closed valves exist, satisfying this requirement. In the provided example, e.g., valves V3 and V4, as well as V1, V2 and V4, or V5 alone etc. See [GLS08b] for an extended analysis.

2.3.6 Common cause failures

Common cause (CC) failures cause multiple channels to fail due to a single event. An example for this effect is a software bug, causing all components with a certain firmware to malfunction at the exact same time (e.g., the year 2000 effect). Another example are transmitters mounted in detached, open-air cabinets in the field (e.g., in sewage plants). In case of an extremely hot summer the temperature in the cabinets rises above the specified limit and thus initiates a common cause failure of all concerned components.

It becomes immediately clear that it is complicated to find CC failures that are not partly systematical failures. In case of, e.g., voltage surges in the energy supply of multiple components, causing them to behave irregularly, it is legitimate to ask why the components were not entirely decoupled or protected from said surges. Systematic failures are not within the scope of the relevant safety standards and thus excluded from all quantitative considerations.

The currently authorized mathematical model is the beta model as described in [DIN02]. It has been under investigation for a long time (consider, e.g., [BL97] from 1997). The IEC 61508 calculation formulas contain terms referring to CC failures derived from the beta model. However, this CC failure model is not uncriticized. The most frequent complaint is that the beta model assumes *all* components in an MooN system fail at the exact same time. It therefore disregards CC failure combinations of two channels in a 2oo3 system. Several extensions to the beta model, covering these shortcomings, are described in, e.g., [D"u10], [BMS10], [HMT06], or - called multiple greek letter model - in [KH96].

The standards suggest heterogeneous instrumentation in order to reduce the fraction of CC failures. The higher the diversity among components, the lower the probability for having multiple devices fail due to a single cause.

2.3.7 Useful component lifetime

The relevant safety standards assume components to have constant failure rates rendering their lifetime exponentially distributed (see sec. 3.3 (p. 32)). This assumption holds only as approximation and only for a specific period of time: the 'useful life' of devices. The real failure rate of safety devices follows more or less the so-called bathtub-curve which can basically be derived from the Hjorth distribution [GC05], [Lit01]. Manufacturers usually deliver their components when they have reached the flat ground of the bathtub. For a certain timespan of 5 to 15 years the devices then are reasonably compliant to the assumption of constant failure rates. Finally, wearout effects become dominant and result in a significant increase of the failure rate. The useful life of the components ends and replacement is required. Devices in SISs therefore have a well-defined mission time assigned which is specified by the respective manufacturer.

Mission time replacement is the only possibility to remove dangerous non-detectable failures (*dn*) from the SIS.

2.3.8 Component repair

As sec. 3.5 (p. 34) will show, unavailability strongly depends on maintenance and thus on component repair processes. Hence, they need to be included into the quantitative considerations. Unfortunately, most repair processes cannot be considered truly random processes. This topic is controversely discussed, e.g., in [Gul03]. Consequently, more or less heavy approximations are necessary to fit said processes into the available calculation models. The usually applied method here is to assume repair times exponentially distributed. This approach is supported and encouraged by various investigations, e.g., [Buk06].

An additional problem of major impact is the number of available repair teams. Obviously, if both components have failed in a 2oo2 SIS, it takes a single repair team longer to restore it back to its functional state, than two teams. The mathematical implementation of this fact is very difficult and controversely discussed as, e.g., [Buk05] shows.

Of greater importance is the fact that two basic classes of failure revelation mechanisms are relevant for SISs in the process industry sector:

- 'Instant repair initiation' is related to automatically detected failures where the repair process is initiated immediately after the occurrence of the failure.
- 'Delayed repair initiation' is related to passive failures that get revealed due to deterministically scheduled proof tests.

Notice that the latter one has a strong deterministic impact that renders it difficult to implement with the available calculation approaches. Both classes are discussed in depth and mathematically in sec. 3.5 (p. 34).

2.3.9 Imperfect component repair

Dangerous undetected failures are typically revealed during a proof test procedure. A common problem with final elements, especially large actuators and valves, is the quality of these proof tests, the proof test coverage (PTC). In certain cases a complete proof test cannot be conducted, since the normal process conditions differ significantly from the conditions during proof tests. If, e.g., a large butterfly valve is supposed to shut the pipe tightly in case of a demand, the correct functionality can only be verified if the valve is dismounted during plant downtime and inspected in a test stand. Otherwise, the stem might be broken in the packaging, suggesting a working component. Hence, a specific fraction of du failures must be assumed dangerous non-detectable (dn). Mathematical investigations of this effect, which is also relevant for the sensor part of an SIS, have been conducted, e.g., in [VPH05], [Buk01], and [GC05].

2.3.10 Maintenance groups

Manual maintenance tests on sensor part side can often be performed by utilizing test-buttons which force a sensor element to temporarily generate an output signal with a configurable signal value in the valid range. By monitoring whether this signal reaches the subsequently allocated SPLC without corruption, possible hidden failures in the whole sensor channel can be revealed. All components in this channel structure that get therefore collectively proof tested form a 'maintenance group'. Apart from test-buttons, further methods are available that allow for a sensor channel test. The proof test of individual devices in a channel is in most cases not possible, although exceptions do exist.

If final element channels are triggered manually via the SPLC for proof testing reasons, the included devices also form a maintenance group. Sometimes the actual final element (e.g., a valve) is blocked in order to perform an online test, i.e., without shutting the process down. As in such cases the valve's safety related functionality cannot be tested, it is to be excluded from the maintenance group containing, e.g., io-card, barrier and solenoid only.

2.4 Derivation of requirements and assumptions

From secs. 2.2 and 2.3 a large number of requirements (abbreviated with 'R') and assumptions ('A') can be derived that will have to be met by the generic approach to be provided in this thesis. This list gives a clear view on the approach's features and restrictions.

The previous sections clearly show that there is a demand for modeling on component layer, as, e.g., sensor signals might get replicated and injected into different channels, thus interconnecting them. This goes beyond the scope of the standards. In IEC 61511, a channel is atomic, i.e., without internal structure and physically totally separated from other channels. An IEC 61511 channel is described by one cumulative failure rate per failure type, representing all components contained in the channel. Therefore, a more detailed view on the SIS should be possible:

R1 Lowest hierarchical modeling layer should be the component layer. A component is assumed to not consist of further subsystems with relevance to SIS applications.

As sensor signals might be splitted, the replicated output of a single sensor might influence, e.g., multiple logic solver functions, and therefore create dependencies between different parts of the SIS.

R2 Component outputs should be connectable to multiple parts of the SIS.

As opposing to IEC 61511, it should not be required to strictly subdivide the whole SIS into sensor part, logic solver, and final element part, each described by an MooN subsystem. As the previous sections have shown, voting functionality might also be implemented in the field in the form of, e.g., field maximum functions or decentralized mini logic solvers as often used in safety relevant PAT.

R3 The subdivision into sensor part, logic solver, and final element part should not be mandatory.

In contrast to the simple component models in IEC 61511 (referring back to IEC 61508), several extensions are required. With respect to maximal flexibility, all failure rates and maintenance parameters should be applicable individually per component, thus allow for an arbitrary degree of diversity.

R4 Failure rates and repair rates should be applicable individually per component.

R5 Each component should be proof testable at arbitrary points in time.

R6 Each component should be replaceable at arbitrary points in time (e.g., due to expiration of its useful lifetime).

Special accuracy should be spent for reproducing the effect of 'maintenance groups', i.e., sets of components which can only be tested together at once (e.g., sensor channels).

R7 Components should be assignable to 'maintenance groups'. The proof test for any component in a maintenance group (e.g., after the repair of an active failure) should result in a proof test for each group member.

A1 It is assumed that a component is proof tested after it has been repaired.

The component's failure behavior should be capable of reproducing the failure modes described in IEC 61511. Extensions with relevance for economic investigations such as active failures are to be considered.

R8 Components should be capable of failing dangerously and safely.

Failures are to be distinguished by the mechanisms that lead to their revelation after a certain period of time.

A2 Dangerous detected (*dd*) failures are assumed to be detected by means of internal diagnostics. The occurrence is communicated using the safety relevant signal (e.g., the 4-20mA current signal with over- or underrange on sensor side). The failure signal requires to reach a device with annunciation capabilities (typically an SPLC) in order to signal the failure to a repair team. The failure signal is not regarded as a sensor trip.

As shown in [Wei09], the consideration of the failure behavior of secondary sensors leads to a significant increase of mathematical complexity which is usually not worth the effort. The additional sensor devices monitoring safety relevant SIS sensors are therefore assumed to be always available.

A3 Dangerous detected failures based on external communication (ddx) are supposed to be detected by additional diagnostic devices with independent communication channels (e.g., additional sensors without direct safety functionality, utilized for detecting dangerous states in primary safety sensors). These secondary monitoring systems are supposed to never fail.

A4 Active failures (a) are assumed to be potentially self-revealing. On sensor side, the false trip signal requires to reach a device with annunciation capabilities (typically an SPLC) in order to get the false channel trip announced. Depending on the applied voting, a false trip of the final element part might be initiated. On final element side, the active failure is detected if it reaches the process connection, i.e., shuts a valve and thus impacts on the process.

A5 The forwarding of failures according to A2 and A4 can only be interrupted by inhibition. A failure is inhibited if a subsequently arranged component (which is responsible for further transmission of the false trip signal) is not capable of performing its intended function.

A6 Dangerous undetected failures (*du*) are not self revealing and remain in the component until a proof test is performed.

A7 Dangerous non-detectable failures (*dn*) are not self revealing and remain in the component until it is replaced, e.g., due to extinction of its useful lifetime.

The recently introduced technique of partial stroke testing enables for advanced proof testing methods. In order to generalize the effect, a 'partial proof test' should be introduced:

R9 Components should be partially proof testable at arbitrary points in time.

A8 Dangerous undetected partial proof test detectable failures (*dup*) are not self revealing and remain in the component until a partial proof test or a regular proof test is performed.

Several further assumptions are required in order to keep the resulting generic approach consistent with the standards, i.e., allow for reproducing comparable unavailability measures.

A9 A component may suffer from only one failure at once.

A10 All components have exponentially distributed lifetimes.

A11 All components have exponentially distributed repairtimes.

A12 The SIS is operated in low demand mode.

A10 requires that electric/electronic/programmable electronic components are operated in their useful lifetime, i.e., while their failure rates remain approximately constant. Components with dedicatedly non-exponential lifetime such as mechanical or electromechanical devices need to be exponentially approximated.

Notice that A11 only refers to the time a repair team spends at a component in order to repair or replace it and conduct a local proof test. The potential queueing time between the component failure and the revelation of said failure is excluded from this consideration.

A12 refers to a demand rate not greater than once per year and not greater than twice the proof test rate as defined in [DIN02] (see also sec. 3.1 (p. 29)).

The aspired flexibility of the ASML approach requires extended features of the utilized voting:

A13 Software voting, e.g., at SPLCs, differs from structural redundancy by its dynamic capabilities. In case of detected failures at arbitrary inputs, a degradation can be performed. Structural redundancy as implemented by, e.g., the arrangement of valves in a pipe network, is always static.

R10 Redundancies should be expressable with more flexibility than the MooN voting scheme, e.g., with a combinatorial voting implementation.

A14 Voters are assumed to have unlimited annunciation capabilities and inform the maintenance crew about every single dangerous detected or active failure at their inputs.

A14 covers both software voting and structural redundancy. Hence, if, e.g., in a 2oo2 final element part, a single valve trips spuriously, this failure is detected due to the impact of the shutting valve on the process.

The last requirement refers to the common cause failure effect:

R11 Common cause failures should be representable.

Notice that this requirement is somewhat weak as it does not demand, e.g., an implementation of the beta model. On the other hand, committing to a specific model removes various degrees of freedom from the engineering process. However, the generic approach to be presented in this work enables for the reproduction of beta model effects as shown in subsec. 5.5.4 (p. 102).

Chapter 3

Basic unavailability principles

3.1 Risk-based SIS design

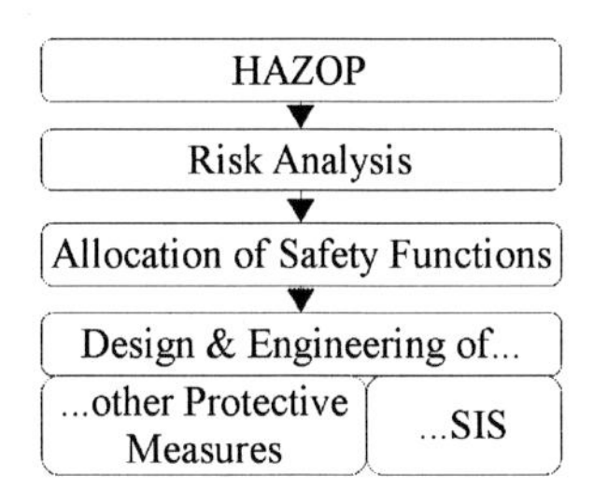

Figure 3.1: Excerpt from the IEC 61511 safety lifecycle

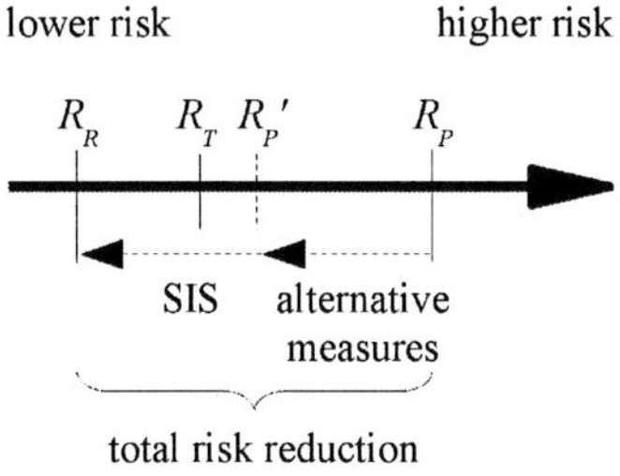

Figure 3.2: Risk graph and risk reduction via SIS

IEC 61511 defines a safety lifecycle for the process industry [DIN05]. This lifecycle provides a complete safety management concept for the identification, planning, implementation, commissioning, maintenance, and decommissioning of protective measures. Although the standard concentrates on automated safety systems as implemented in a SIS, suitable specification is given for alternative safety systems such as organizational or mitigation measures. An important part of the safety lifecycle deals with the assessment of functional safety. The considered part is shown as an excerpt in fig. 3.1. A safety expert team conducts a hazard and operability (HAZOP) investigation with the intention of identifying all potential hazards that may emanate from a process. Typically, this is performed by segmenting the process P&ID into individual apparatuses and discussing a standardized list of buzzwords such as 'too high', 'too low', 'too much' etc. with each relevant characteristic of the process if treated by the considered apparatus. This leads to potentially dangerous combinations like 'high pressure in a vessel' which have to be investigated further in the next step, the risk analysis. Here, a process risk R_P is assigned to each individual hazard from the hazard list. Depending on whether this initial risk is higher than a hazard individual tolerable risk R_T (see risk graph in fig. 3.2), at least one protective measure is required. During the allocation phase, a suitable combination of measures is defined that in sum reduce the process risk to an acceptable residual risk R_R far enough to the left of the tolerable risk. This reduction can be accomplished by applying a SIS or alternative measures such as housings, organizational measures, or evacuation plans [1]. Since 2003, the safety standards for the process industry demand to conduct this risk assessment at least semi-quantitatively. The authorized risk assessment methods such as the layers of protection analysis (LOPA) or the risk graph result in concrete quantitative safety requirements for the probably required SIF. The chosen parameter for denoting the quality of a SIF is its average probability of failure on demand (PFD_{avg}), i.e., its average safety related unavailability. The standard distinguishes four categories of quality, the safety integrity levels (SIL). As can be seen from table 3.1, the quality levels

Table 3.1: Safety integrity level (SIL)

SIL	PFD_{avg}
1	$\geq 0.01 ... < 0.1$
2	$\geq 0.001 ... < 0.01$
3	$\geq 0.0001 ... < 0.001$
4	$\geq 0.00001 ... < 0.0001$

are logarithmically staggered. Typically, the result of the risk assessment phase is, e.g., the specification of a SIF that has to ensure the pressure relieving for a vessel in case of a demand with SIL2 quality. The implementation of this SIF, the SIS, may solve this task by using, e.g., overpressure sensors, and connecting these to relieve valves via an SPLC.

Following [Lit98], the initial process risk R_P can be defined as the product of the demand frequency f_D (the frequency of the process getting into a dangerous state) and the expected value of the potentially caused damage's severity $\mathrm{E}(S)$.

$$R_P = f_D \cdot \mathrm{E}(S). \tag{3.1}$$

This fundamental equation assumes that a demand immediately leads to the undesired threat with expected severity $\mathrm{E}(S)$ if no protective measures such as a SIS is present. If $\mathrm{E}(S)$ is given in, e.g., Euros, the process risk R_P can be interpreted as the expected amount of Euros per year that have to be spent in order to cover the damage caused by the considered process. The intention of SISs and alternative protective measures is to lower this hazard frequency. The undesired event only occurs if a demand is pending and all protection layers fail.

After the transition from frequencies to probabilities (related to equal time intervals), eq. 3.1 reads as

$$R_P = \mathrm{P}(D) \cdot \mathrm{E}(S), \tag{3.2}$$

[1] The exemplary HAZOP for fig. 3.2 delivered that a combination of SIS and alternative measures is required, as only the combination of both results in a residual risk R_R below the tolerable risk R_T.

where D represents the event 'a demand occurs'. After taking into consideration all alternative protection measures, a process risk R'_P remains, that has to be covered by a SIS solely. Following [DLN08], the effect of this SIS can then be considered via

$$R_R = \mathrm{P}(F \cap D') \cdot \mathrm{E}(S),$$

where R_R is the residual risk, i.e., the remaining risk after the application of the SIS, F is the event 'a failure in the SIS prevents it from performing its intended safety function', and D' is the event 'a demand occurs and all alternative protection measures fail'. This can generally be expressed as

$$R_R = \mathrm{P}(F \mid D')\,\mathrm{P}(D') \cdot \mathrm{E}(S),$$

using conditional probabilities. In case of strictly separated protection layers, F and D' are stochastically independent from each other and thus $\mathrm{P}(F \mid D') = \mathrm{P}(F)$, resulting in

$$R_R = \mathrm{P}(F) \cdot \underbrace{\mathrm{P}(D') \cdot \mathrm{E}(S)}_{R'_P}. \tag{3.3}$$

It is obvious that the probability $\mathrm{P}(F)$ is time variant since the probability of failure of a SIS strongly depends on lifetime distributions of the underlying components as well as maintenance strategies. In order to simplify the mathematical terms, the safety standards demand that eq. 3.3 is satisfied on average. Thus, $\mathrm{P}(F)$ can be replaced by the average probability of failure on demand, i.e., the safety related unavailability as introduced by IEC 61508:

$$R_R = PFD_{avg} \cdot R'_P.$$

It is quite important to point out again, that the provided derivation depends on the assumption that the events F and D' are stochastically independent from each other. As shown in, e.g., subsec. 2.2.3 (p. 17), modern SISs may violate this precondition. Special treatment of such systems is required.

Notice that the provided considerations hold for the so-called low demand mode according to A12 only. It refers to a demand rate not greater than once per year and not greater than twice the proof test rate as defined in [DIN02]. Otherwise, the probability of failure per hour (PFH) is to be calculated instead of the PFD. A detailed differentiation of both terms is discussed in, e.g., [IDRS08], and [BvHS09].

3.2 Cost-effective SIS design

In contrast to the safety related PFD_{avg}, operational characteristics represent the probability of having the SIS in completely or partially tripped state while no demand is pending. Plant operators try to avoid such spurious trips of their SISs, as unintended shutdowns of the systems lead to horrendous costs of idleness. Moreover, each shutdown implies a new startup phase which also bears additional risks which are often not covered in the risk assessment phase of the safety lifecycle. Therefore, plant operators always put large effort into finding a SIS design leading to a safety loop which is compliant to the safety standards and - at the same time - offers maximal operational availability.

It is easily possible to design SISs with nearly identical PFD_{avg}, but completely different operational unavailability. Standard 1oo2 and 2oo3 redundant SIS parts have approximately the same averaged safety related unavailability, but differ significantly in their operational unavailability (assuming identical components and maintenance). The spurious trip rate, i.e., the frequency of occurrence of spurious trips is around one decade higher for the 1oo2 solution. If an emergency shutdown of the considered plant would lead to time consuming cleanup and startup processes, the higher investment costs for the 2oo3 SIS can be neglected. Various methods have been developed to find cost optimal SIS designs under PFD_{avg} constraints, e.g., [ML11], [TEMT09a] or [TEMT09b].

As the safety standards do not consider operational unavailability effects, this characteristic will be defined here, following [HR05]:

Definition 3.1 (Average probability of fail-safe). *The PFS_{avg} (average probability of fail-safe) is the natural counterpart to the PFD_{avg} and specifies the average probability of meeting a SIS in a state where it has unintendedly executed its assigned SIF due to the spurious trip of an integral component.*

This means, that, e.g., in a 2oo2 final element part, both valves must have shut. A system with $PFS_{avg} = 0.01$ and process dependent idle costs of 1000EUR per hour result in $87,600$EUR per year if the SIS is considered separately from other safety loops (1 year equals 8760 hours).

3.3 Unreliability

It is important to stress the differences and similarities of unreliability, maintenance, and unavailability. All these terms are frequently used in this work and thus their meaning has to be clarified within the required scope. When talking about the unreliability of a component or a system, the function

$$\mathrm{F}(t) = \mathrm{P}(TTF \leq t) \tag{3.4}$$

is referred to [Lit01]. It denotes the probability that the random variable *time-to-failure* (TTF) for the considered component or system has a value lower or equal t. In other words: a component or system that started at time $t = 0$ in its functional state will have failed at $t = T$ with probability $\mathrm{F}(T)$. Due to this effect, the function $\mathrm{F}(t)$ is also called the *point-unreliability*. It is important to emphasize on the fact, that unreliability considers only *repair-failure processes* (RF processes, see [KH96]). These processes represent components starting in their functional state (repaired) failing afterwards, and remaining in their failed state forever.

Using the term 'unreliability' does *not* allow for predictions on whether a SIS is functional at $t = 12000h$: a safety instrumented system is a maintained system and thus not permanently remaining in its unfunctional state. Therefore, the only possible deduction based on unreliability is the probability for the SIS to fail *at least once* until $t = 12000h$. Notice the slight difference of information content.

Unreliability functions are cumulative density functions (CDFs) and have thus typical characteristics. The *steady-state unreliability* is

$$\lim_{t \to \infty} \mathrm{F}(t) = 1,$$

i.e., for $t \to \infty$ every system or component will fail definitely. Sometimes it is of interest to get an average unreliability over a certain observation period T_{obs} starting from $t = 0$ according to

$$\overline{\mathrm{F}}(T_{obs}) = \mathrm{F}_{avg}(T_{obs}) = \frac{1}{T_{obs}} \int_{t=0}^{T_{obs}} \mathrm{F}(t)dt.$$

This parameter is of interest if, e.g., a large population of systems or components of the same type, are under consideration. If the individuals have equally distributed origins of their time axes and get replaced after T_{obs}, then a randomly picked individual at any point in time will be in its failed state with probability $\overline{\mathrm{F}}(T_{obs})$.

The exponentially distributed RF process as the most important one with respect to this thesis (among RF processes with different distributions), has the unreliability

$$\mathrm{F}(t) = 1 - \exp(-\lambda t), \tag{3.5}$$

with failure rate λ and specific characteristics

$$\lim_{t \to \infty} \mathrm{F}(t) = \lim_{t \to \infty} \left(1 - \underbrace{\exp(-\lambda t)}_{\to 0} \right) = 1,$$
$$\overline{\mathrm{F}}(T_{obs}) = 1 + \frac{\exp(-\lambda T_{obs}) - 1}{\lambda T_{obs}}.$$

3.4 Maintenance

Analog to RF processes, the random variable *time-to-restoration* (TTR) of a failed system can be introduced in order to describe the time a component spends in its failed state until it is functional again:

$$\mathrm{M}(t) = \mathrm{P}(TTR \leq t) \tag{3.6}$$

In contrast to the RF processes, these *failure-repair* processes (FR processes, see [KH96]) are much more complicated to derive. They are of course CDFs of the random variable TTR, but often the explicit probability distribution cannot be determined as the chosen maintenance strategy significantly impacts on the problem complexity. The major problem is that typically multiple subprocesses impact on the TTR: failure detection time, waiting for spare parts, the actual repair time etc. All of these subprocesses could be modeled separately and superimposed to form the TTR. In practice, only the most important subprocesses with respect to the chosen maintenance strategy are considered and the rest is neglected in order to keep things simple.

For this work, two basic maintenance strategies are of great importance:

- Instant Repair Initiation
- Delayed Repair Initiation

The first one refers to the maintenance of, e.g., dangerous detected (*dd*) failures. This class of failures is self revealing and leads to rapid restoration. Here, the detection time for the failure is neglected and the related FR process can be modeled with a simple repair time distribution such as an exponential or - more realistic - a lognormal distribution.

Delayed repair initiation refers to, e.g., dangerous undetected (*du*) failures. These failures are not immediately detected after the failure occurrence. Instead, they remain unrevealed until the next proof test is conducted for the considered component. Then the actual repair is initiated. The queuing time between the failure event and the upcoming proof test is complicated to derive as it depends on the actual time of the failure occurrence (a random variable defined by the related RF process) as well as on a regularly conducted proof test (a non-stochastic process). Compared to the queuing time, the repair duration is negligible.

By superimposing RF and FR process, an RFR (*repair-failure-repair*, see [KH96]) cycle results. As the explicit points in time at which a component fails or gets repaired are unknown, only stochastical conclusions concerning the functionality of the component at arbitrary points in time can be drawn. This is accomplished by advancing to unavailability calculation, where the combination of failures and maintenance is mathematically conducted.

3.5 Unavailability

Following [RH04], unavailability basically means

$$\mathrm{U}(t) = \mathrm{P}(\text{system is not functional at time } t).$$

This *point-unavailability* denotes the probability that a system or component that started at time $t = 0$ in its functional state will be in its failed state at $t = T$ with probability $\mathrm{U}(T)$. It is important to emphasize on the fact, that an unknown number of failure and repair processes may have occurred for $0 \leq t \leq T$. Unavailability is based on an underlying RFR cycle of failure and restoration processes. One cycle period always consists of an RF process describing the failure mechanisms, as well as of an FR process, describing the restoration of a component's or system's functionality. Therefore, the RFR processes cycle time (an individual random variable) can be expressed as the sum of the underlying random variables $TTF+TTR$. FR and RF processes can be individually described by the previously introduced unreliability and maintenance functions according to subsecs. 3.3 and 3.4.

Unavailability functions are generally not CDFs as the *steady-state-unavailability* shows:

$$\lim_{t\to\infty} \mathrm{U}(t) = \lim_{t\to\infty} \frac{\mathrm{DT}(t)}{\mathrm{DT}(t) + \mathrm{UT}(t)} = \frac{MDT}{MDT + MUT} \neq 1. \tag{3.7}$$

Here, $\mathrm{DT}(t)$ is the total downtime over time, i.e., the component's or system's time spent in its failed state from its birth up to t. $\mathrm{UT}(t) = t - \mathrm{DT}(t)$ is the item's uptime. The mean down time (MDT) and the mean up time (MUT) are well known parameters that can be estimated from statistics and have been explained in, e.g., [Lit98], and [Rom05].

According to the extended characteristics introduced with unreliability, the *average unavailability* over an observation period T_{obs} can be defined as

$$\overline{\mathrm{U}}(T_{obs}) = \mathrm{U}_{avg}(T_{obs}) = \frac{1}{T_{obs}} \int_{t=0}^{T_{obs}} \mathrm{U}(t)dt. \tag{3.8}$$

This parameter is of great importance as the average probability of failure on demand (PFD_{avg}) is an average unavailability [DIN02]. The meaning of an average PFD should be clarified, using a descriptive example:

> Given a population of 100 identical SISs with $PFD_{avg} = 0.1$ in equally distributed states of their lifespans of length T_{obs}. Then an average of 10 SISs is safety relatedly unavailable at any arbitrary point in time.

The basic approach towards unavailability calculation is as follows: be a single component subject to an RFR cycle with RF process described by the unreliability density $\mathrm{f}(t) = \frac{d\mathrm{F}(t)}{dt}$ according to eq. 3.4 (p. 32) and an FR process given as restoration time probability density function $\mathrm{m}(t) = \frac{d\mathrm{M}(t)}{dt}$ according to eq. 3.6 (p. 33). Then, according to [KH96], its unavailability can be derived from

$$\mathrm{U}(t) = \int_0^t \left[w(u) - v(u)\right] du, \tag{3.9}$$

where $w(u)$ and $v(u)$ can be obtained from

$$w(t) = \mathrm{f}(t) + \int_0^t \mathrm{f}(t-u)v(u)du, v(t) = \int_0^t \mathrm{m}(t-u)w(u)du.$$

It is generally difficult to set up the correct unavailability function for a given RFR process. Said difficulties derive from potential variations in the FR part of the RFR cycle as anticipated in subsec. 3.4 (p. 33). In literature, a variety of such specific scenarios is analyzed and suitable unavailability functions are available. The subsequent paragraphs outline the two most important special cases of these RFR processes that will be applied in several variations throughout this thesis. The second one - RFR processes with delayed repair initiation - is presented in three variations, as all three are of equal importance.

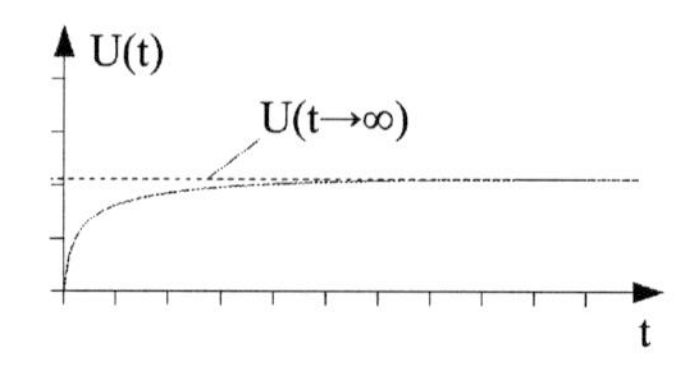

Figure 3.3: Unavailability for RFR processes with instant repair initiation

3.5.1 Unavailability for RFR-IRI processes

The most important characteristics of RFR processes with instant repair initiation (RFR-IRI processes):

- enables for calculation of unavailability for components subject to, e.g., *dd* failures
- the TTF of the RF process is assumed to be exponentially distributed with failure rate λ
- the TTR of the FR process consists of the pure repair duration only and is assumed to be exponentially distributed with repair rate μ
- failure detection time and further queuing processes are neglected

Parameterizing eq. 3.9 with the given PDFs of the RF and FR process, the solution can be obtained via Laplace transformation as shown in [KH96]:

$$\mathrm{U}(t) = \frac{\lambda}{\lambda + \mu} \left[1 - exp\left(-\left(\lambda + \mu\right) t\right)\right]. \tag{3.10}$$

From this point-unavailability, it is easy to derive steady-state and average unavailability:

$$\lim_{t \to \infty} \mathrm{U}(t) = \lim_{t \to \infty} \frac{\lambda}{\lambda + \mu} \left[1 - \underbrace{exp\left(-\left(\lambda + \mu\right) t\right)}_{\to 0}\right] = \frac{\lambda}{\lambda + \mu}, \tag{3.11}$$

$$\overline{\mathrm{U}}(T_{obs}) = \frac{\lambda}{\lambda + \mu} + \frac{\lambda}{\left(\lambda + \mu\right)^2 T_{obs}} \left[exp\left(-(\lambda + \mu) T_{obs}\right) - 1\right].$$

$\mathrm{U}(t)$ is shown in fig. 3.3. The steady-state unavailability is denoted as $\mathrm{U}(t \to \infty)$. Notice that $\lim_{t \to \infty} \mathrm{U}(t) \approx \overline{\mathrm{U}}(T_{obs})$ for large T_{obs}. This property is exploited heavily in system unavailability assessment. Subsequent sections will refer to this effect.

3.5.2 Unavailability for RFR-DRI processes - type A

For RFR cycles with delayed repair initiation (RFR-DRI processes), three variations are commonly used. They differ by their degree of accuracy in reproducing the influence of deterministic periodic inspection on the considered components. A consistent and closed derivation for all three approaches in the context of unavailability calculation for the process industry is not known to the author. It will therefore be done subsequently with the intention of preparing for their respective implementation in different unavailability modeling techniques (to be presented in the subsequent chapter 4). The most important characteristics of type A RFR-DRI process:

- enables for calculation of unavailability for components subject to, e.g., *du* failures
- the TTF of the RF process is assumed to be exponentially distributed with failure rate λ
- the TTR of the FR process consists of the queuing time from the failure occurrence until the upcoming periodic proof test at nT_I
- actual repair time and further processes are neglected
- the proof test interval T_I is assumed to be constant

The basic idea is to approximate this complex RFR process with a more simple RFR-IRI process as previously introduced in subsec. 3.5.1 (p. 36). A component with exponentially

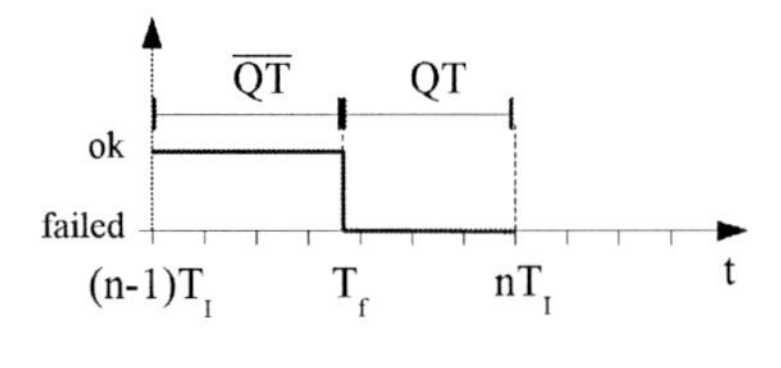

Figure 3.4: Derivation of the FR process of an RFR-DRI process of type A

distributed lifetime $\mathrm{F}(t) = 1 - \exp(-\lambda t)$ is periodically tested each nT_I. In case of a failure at $t = nT_I$ it is repaired instantly. The component's lifetime thus ends after a time $\overline{QT}$ inbetween two subsequent proof tests. This time is a random variable with PDF $f'_{\overline{QT}}(t)$ and matching CDF $F'_{\overline{QT}}(t)$. The expected queuing time QT, i.e., the time from the failure occurrence at $T_f = (n-1)T_I + \overline{QT}$ until the upcoming proof test at nT_I is

$$\mathrm{E}(QT) = T_I - \frac{1}{\lambda} + \frac{T_I}{\exp(\lambda T_I) - 1}. \tag{3.12}$$

See sec. 10.1 (p. 167), as well as [Sch76] and [Sch80], for an extended derivation. By expanding eq. 3.12 into a MacLaurin series (see sec. 10.2 (p. 168)) using three summands,

$$\mathrm{E}(QT) \approx T_I - \frac{\frac{1}{2}T_I}{1 + \frac{1}{2}\lambda T_I} \overset{\lambda T_I \ll 1}{\approx} \frac{1}{2}T_I$$

follows as approximation. In exponential repair time processes, the expected value of the related CDF, i.e., the expected repair time $E(TTR)$, corresponds to μ^{-1}. Treating the queuing time QT as an exponentially distributed random variable (which it is not), the related characteristic parameter μ can be chosen as

$$\mu = \frac{1}{E(QT)} \approx \frac{2}{T_I}.$$

The correctly modeled RFR-DRI process can thus finally be approximated with a more simple RFR-IRI process according to eq. 3.10 (p. 36):

$$\begin{aligned} U(t) &\approx \frac{\lambda}{\lambda + \frac{2}{T_I}} \left[1 - exp\left(-\left(\lambda + \frac{2}{T_I}\right)t\right)\right] \\ \lim_{t\to\infty} U(t) &\approx \frac{\lambda}{\lambda + \frac{2}{T_I}} \overset{\lambda T_I \ll 1}{\approx} \frac{1}{2}\lambda T_I. \end{aligned} \tag{3.13}$$

As this unavailability function has lost the relation to absolute time and can only reproduce the expected value of the correct CDF, the steady-state unavailability consideration alone makes sense. Average and point-unavailability are not suitable of being used in extensive analyses. Although subject to that many simplifications and approximations, this approach is utilized commonly, e.g., in markov (sec. 4.4 (p. 50)), and fault tree analysis (sec. 4.3 (p. 45)). The expected queuing time $E(Q) \approx \frac{1}{2}T_I$ is extensively used in the derivation of the IEC 61508 standard calculation formulas. However, as already mentioned in subsec. 3.5.1 (p. 36), the average unavailability is similar to the steady-state unavailability for large T_I, rendering this approximation admissible for most applications.

3.5.3 Unavailability for RFR-DRI processes - type B

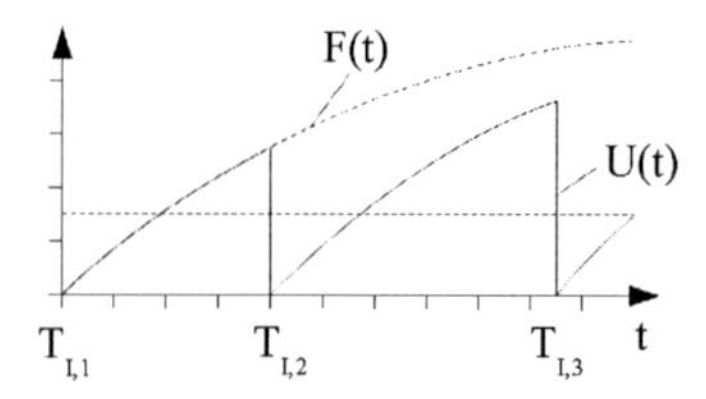

Figure 3.5: Unavailability for type B RFR processes with delayed repair initiation

The most important characteristics of this RFR-DRI process:

- enables for calculation of unavailability for components subject to, e.g., *du* failures
- the TTF of the RF process is assumed to be exponentially distributed with failure rate λ

- the TTR of the FR process consists of the queuing time starting from the failure occurrence until the upcoming proof tests at $T_{I,1}, T_{I,2}, \ldots$
- actual repair time and further processes are neglected
- the adhoc information $U(nT^{+}_{I,n}) = 0$ (after each proof test the component is definitely functional) is utilized
- The unavailability function keeps its relation to absolute time (compare 3.5.2) and thus enables for point-unavailability calculation

The basic idea is to skip the derivation via basic approach from eq. 3.9 and use the ad-hoc information $U(nT^{+}_{I,n}) = 0$ to directly construct the unavailability $U(t)$. If restoration occurs only at predetermined points in time and is instantly completed, the time span between these points is totally dominated by the RF process, i.e., the component's unreliability. A suitable unavailability function $\mathrm{U}(t)$ for m proof tests can thus easily be constructed according to

$$\mathrm{U}(t) = \sum_{i=1}^{m-1} \mathrm{F}(t - T_{I,i}) \left(\sigma\left(t - T_{I,i}\right) - \sigma\left(t - T_{I,i+1}\right)\right),$$

see also fig. 3.5. At proof tests $T_{I,1}, T_{I,2}$, and $T_{I,3}$ the component's unavailability is reset back to zero instantly as the duration of the repair is neglected and results in a fully functional component. It is obvious, that steady-state, and average unavailability calculation is possible but not feasible. The resulting expressions are very unhandy and usually not applicable in further considerations.

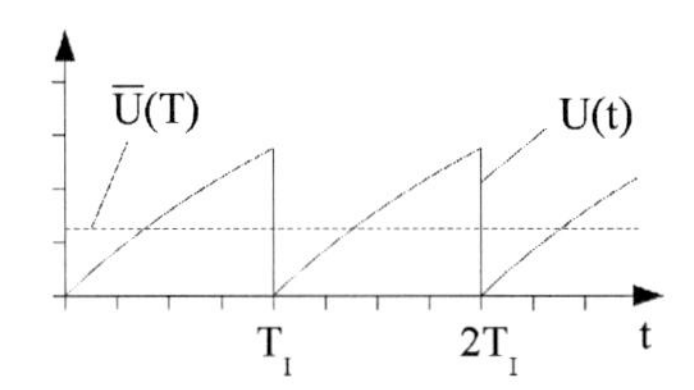

Figure 3.6: Periodic unavailability for type B RFR-DRI processes

Instead, one important further simplification that enables for a brief closed mathematical expression assumes equidistant proof test times $T_{I,1} = T_{I,2} = \ldots = T_I$ (see fig. 3.6). Then the component's unavailability is periodic with length T_I and can be expressed as

$$\mathrm{U}(t) = \mathrm{F}(t \bmod T_I).$$

Here, 'mod' is the modulo operation. Due to the periodicity, it is easy to derive

$$\overline{\mathrm{U}}(T_I) = \frac{1}{T_I} \int_0^{T_I} \mathrm{F}(t)dt = 1 + \frac{\exp\left(-\lambda T_I\right) - 1}{\lambda T_I} = \lim_{t \to \infty} \overline{\mathrm{U}}(t). \tag{3.14}$$

Expanding eq. 3.14 into a MacLaurin series, delivers the exact same result as eq. 3.13 (p. 38) for the RFR-DRI process of type A.

3.5.4 Unavailability for RFR-DRI processes - type C

The most important characteristics of this RFR-DRI process:

- enables for calculation of unavailability for components subject to, e.g., *du* failures
- the TTF of the RF process is assumed to be exponentially distributed with failure rate λ
- the TTR of the FR process consists of the queuing time Q starting from the failure occurrence until the upcoming proof test at nT_I as well as an exponentially distributed repair time with repair rate μ_R

In contrast to the RFR-DRI processes of type A and B, the TTR for this process consists of two independent random variables, the queuing time Q as defined in subsec. 3.5.2 (p. 36), and the time-to-repair $TTRep$, such that

$$TTR = Q + TTRep.$$

It is easy to extend the RFR-DRI type A approach adequately by setting

$$\mu = \frac{1}{\mathrm{E}(Q) + \mathrm{E}(TTRep)} \approx \frac{1}{\frac{T_I}{2} + \frac{1}{\mu_R}}.$$

This can easily be derived from the well known fact (see, e.g., [Lit01]) that for arbitrary two random variables X and Y the following holds: $\mathrm{E}(X+Y) = \mathrm{E}(X) + \mathrm{E}(Y)$. The RFR-DRI process as described here is closer to real RFR cycles than the type A process, as it enables for an actual consideration of repair times in addition to the queuing time. An analogical extension for the type B approach is not trivial and results in extended mathematical effort. Section 4.4 introducing markov models will provide another elegant way of implementing this RFR-DRI type C process mathematically.

3.5.5 Unmaintained systems

Unmaintained systems always fall back to unreliability functions. This can be shown by zeroizing the repair terms in arbitrary unavailability functions. Equation eq. 3.10 (p. 36) describing an RFR-IRI process, where RF as well as FR process are exponentially distributed, may serve as an example. By setting $\mu = 0$ (i.e., no repair),

$$U_{\mu=0}(t) = \frac{\lambda}{\lambda + 0}\left[1 - exp\left(-\left(\lambda + 0\right)t\right)\right] = 1 - exp\left(-\lambda t\right) = \mathrm{F}(t).$$

follows. The same result can be obtained from the more general approach according to eq. 3.7 (p. 35). For arbitrary unmaintained systems

$$\lim_{t\to\infty} \mathrm{U}(t) = \lim_{t\to\infty} \frac{\overbrace{\mathrm{DT}(t)}^{\to\infty}}{\underbrace{\mathrm{DT}(t)}_{\to\infty} + \underbrace{\mathrm{UT}(t)}_{\to const}} = 1 = \lim_{t\to\infty} \mathrm{F}(t). \tag{3.15}$$

holds, as the downtime approaches infinity for increasing time since after the first failure no repair is applied.

Chapter 4

Unavailability calculation methods for modern SISs

4.1 Overview

Various approaches for the calculation of important safety characteristics such as the PFD do exist. They are used to treat larger systems, where lots of components with individual and sometimes interdependent RFR cycles are interconnected with each other, building up channel structures and redundancies. Most approaches start from a graphical system model, hiding the mathematical complexity of the problem and allowing for the representation of as many relevant effects as possible. From this graphical representation of the SIS, a mathematical calculation model is derived, and finally well-established algorithms enable for the retrieval of the desired characteristics. The available methods differ in complexity and the provided effects reflectable by the model. Generally, two basic types of unavailability calculation approaches can be distinguished that are authorized for PFD calculation by IEC 61508:

- Combinatorial approaches and
- State based approaches

Both types have prominent representatives that have been used in unavailability calculation for a long time.

Combinatorial approaches usually provide a graphical representation of the utilized components of a SIS and their interconnections. It is quite easy to recognize the physical structure of the real system in the SIS model. The related calculation model is typically based on combinatorial terms representing probabilistic relations between the individual components. In this section, two of the most popular combinatorial calculation approaches are introduced - reliability block diagrams (RBD) and fault trees (FT).

In contrast to combinatorial approaches, state based methods commonly utilize some kind of automaton in order to represent the internal state of a SIS at any point in time. Since automata implicitly contain time variance in form of the transitions between states, they are out of the box more suited to model complex dynamic behavior. On the other hand, state based approaches usually result in very large system models. Most important representatives for state based approaches are markov models and monte carlo simulations.

This section is meant to provide an overview of the general approaches as usually applied in the process industry sector. All presented methods are applied in recent research activities worldwide and come in uncountable variations and adaptations for various specific problems. An overview within the scope of IEC 61511 and IEC 61511 provide, e.g., [B"o06], [KH96], [Rou04], [Gob00], and [RvdB02]. If variations are available, the most important ones with respect to this thesis are picked. The graphical layer is introduced first for each approach. Afterwards, the related mathematical representation is provided. Depending on the utilized RFR cycle types, several algorithms are available that simplify the mathematical complexity of the solution. Therefore, the implementation of the most important maintenance strategies as introduced in sec. 3.5 (p. 34) is presented for each approach.

The new generic approach as defined in chaps. 3 to 5 refers to several of these standard approaches as it extends their capabilities and combines them with completely new methods with the intention of maximizing the amount of treatable modern SISs.

4.2 Reliability block diagrams

4.2.1 General concept

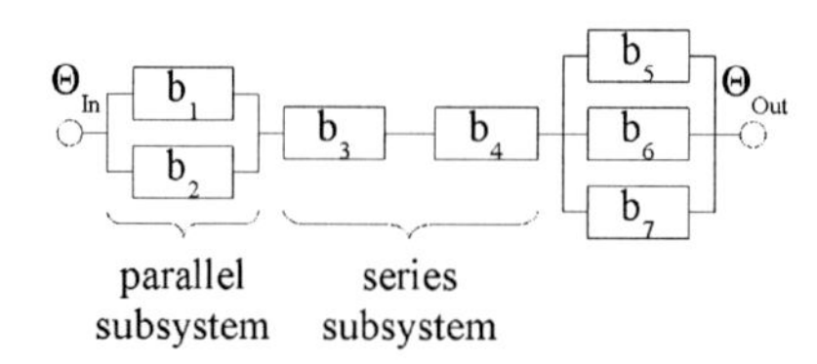

Figure 4.1: RBD example

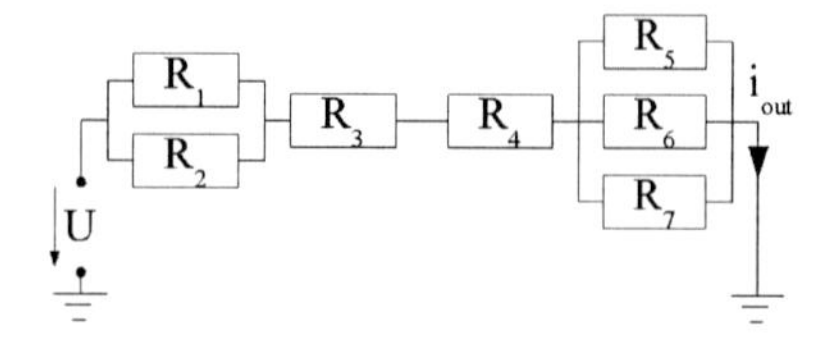

Figure 4.2: Resistor network equivalent to the RBD in fig. 4.1

As can be derived from the name of this approach, the traditional reliability block diagrams (RBD) are meant for reliability calculations. A generalization towards (un-) availability is straight forward and preferred, as all required target characteristics such as the PFD are unavailability measures. It is of course possible to obtain unreliabilities by removing all maintenance terms from the unavailability functions as shown in subsec. 3.5.5 (p. 40).

RBDs are bi-terminated directed graphs that consist of a set of blocks $\mathcal{B}$ as well as a set of signal flow paths, the edges $\mathcal{E}$:

$$\mathcal{RBD}\left(\mathcal{B}, \mathcal{E}, \Theta_{In}, \Theta_{Out}\right).$$

Each block $b_i \in \mathcal{B}$ represents a physical device which contributes to the system's unavailability with its RFR cycle information. In the RBD graph each block is drawn as a vertex. The graph's edges are valid signal flow paths according to

$$\mathcal{E} \subseteq ((\{\Theta_{In}\} \times \mathcal{B}) \cup (\mathcal{B} \times \mathcal{B}) \cup (\mathcal{B} \times \{\Theta_{Out}\}))$$

and may refer to physical as well as to software signal transfer. Although they are defined as directed edges, the direction of communication is irrelevant for the standard unavailability calculation process. Moreover, the directive information for an arbitrary edge can implicitly be derived from the location of Θ_{In} and Θ_{Out} in the graph. An RBD system is considered available as long as a 'path' of available components exists, that interconnects the input terminal Θ_{In} with the output terminal Θ_{Out} (see fig. 4.1). This is to be concluded in a probabilistic way, since components are defined via their unavailability functions only. The functionality of RBDs can be explained using an analogy from the area of electrical engineering. Thus, an RBD can be interpreted as a potential causing a current through a resistor network (see fig. 4.2), where

- electrical wires represent RBD signal flow paths,
- short circuited resistors ($R = 0$) represent available components,
- open circuit resistors ($R \rightarrow \infty$) represent unavailable components.

If the output current $i_{out} \neq 0$, the represented SIS is available, otherwise it is not. The related calculation models lead to probabilistic combinatorial equations. Simple redundancies (1ooN) are expressed as pure parallel block subsystems. Necessarily required block structures are arranged in series. Standard MooN voting schemes such as 2oo3 are

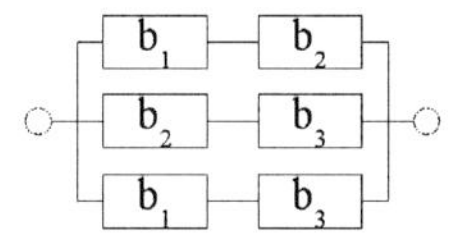

Figure 4.3: Implementation of a 2oo3 redundancy in an RBD

modeled by instantiating a concrete component several times and assigning it to multiple paths in parallel structures as can be seen from fig. 4.3. Therefore

$$|\{b_i \mid b_i \in \mathcal{B} \wedge \mathrm{U}_i = \mathrm{U}_{spec}\}| \geq 1,$$

where U_{spec} is an arbitrary but specified unavailability function, and U_i is the unavailability function assigned to the i-th block.

4.2.2 Calculation model

Assuming stochastically independent component RFR cycles, the RBD enables for the derivation of a system's unavailability function $\mathrm{U}_{SIS}(t)$ which is a function of the individual component's unavailability functions U_i. It is easy to derive

$$\mathrm{U}_{series}(t) = \mathrm{f}(\mathrm{U}_1(t), \ldots, \mathrm{U}_i(t), \ldots, \mathrm{U}_n(t)) = 1 - \prod_{i=1}^{n} (1 - \mathrm{U}_i(t)) \tag{4.1}$$

$$\mathrm{U}_{parallel}(t) = \mathrm{f}(\mathrm{U}_1(t), \ldots, \mathrm{U}_i(t), \ldots, \mathrm{U}_n(t)) = \prod_{i=1}^{n} \mathrm{U}_i(t) \tag{4.2}$$

for series or parallel structures with n components.

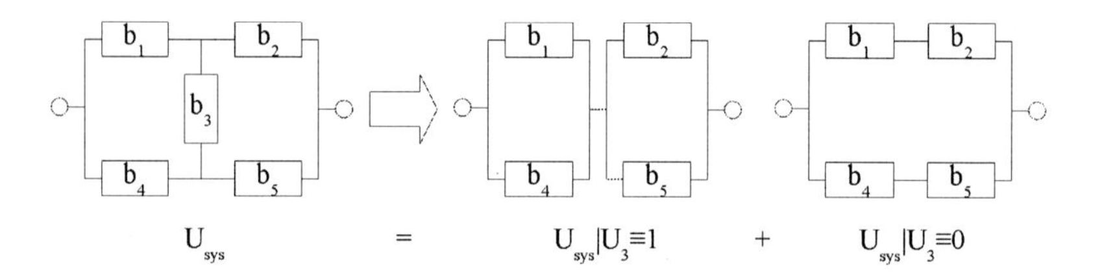

Figure 4.4: Decomposition of hybrid structures in RBDs

According to the sum of products principle as introduced, e.g., in [Har09], the decomposition of hybrid structures is possible:

$$\begin{aligned}\mathrm{U}_{SIS}(t) &= \mathrm{f}(\mathrm{U}_1(t), \ldots, \mathrm{U}_i(t), \ldots, \mathrm{U}_n(t)) \\ &= \mathrm{U}_i(t)\,\mathrm{f}(\mathrm{U}_1(t), \ldots, 1, \ldots, \mathrm{U}_n(t)) + (1 - \mathrm{U}_i(t))\,\mathrm{f}(\mathrm{U}_1(t), \ldots, 0, \ldots, \mathrm{U}_n(t)).\end{aligned}$$

For the bridge example from fig. 4.4 this leads to the first summand describing a series of two parallel structures ($b_1 \parallel b_4$ in series with $b_2 \parallel b_5$), and the second one a parallel structure of two series ($b_1 - b_2$ parallel $b_4 - b_5$) according to

$$\begin{aligned}\mathrm{U}_{4.4}(t) = \mathrm{U}_3(t) & \underbrace{\left[1 - \left[\left(1 - \underbrace{\mathrm{U}_1(t)\,\mathrm{U}_4(t)}_{b_1\|b_4}\right)\left(1 - \underbrace{\mathrm{U}_2(t)\,\mathrm{U}_5(t)}_{b_2\|b_5}\right)\right]\right]}_{(b_1\|b_4)-(b_2\|b_5)} \\ + (1 - \mathrm{U}_3(t)) & \underbrace{\left[\underbrace{1 - [(1 - \mathrm{U}_1(t))\,(1 - \mathrm{U}_2(t))]}_{b_1-b_2}\right]\left[\underbrace{1 - [(1 - \mathrm{U}_4(t))\,(1 - \mathrm{U}_5(t))]}_{b_4-b_5}\right]}_{(b_1-b_2)\|(b_4-b_5)}\end{aligned}$$

and thus

$$\begin{aligned}
\mathrm{U}_{4.4}(t) = \quad & \mathrm{U}_1(t)\,\mathrm{U}_5(t) + \mathrm{U}_2(t)\,\mathrm{U}_4(t) + \mathrm{U}_2(t)\,\mathrm{U}_5(t) + \mathrm{U}_1(t)\,\mathrm{U}_4(t)\,\mathrm{U}_4(t) \\
& - \mathrm{U}_1(t)\,\mathrm{U}_4(t)\,\mathrm{U}_5(t) - \mathrm{U}_1(t)\,\mathrm{U}_2(t)\,\mathrm{U}_5(t) - \mathrm{U}_1(t)\,\mathrm{U}_2(t)\,\mathrm{U}_4(t) \\
& - \mathrm{U}_1(t)\,\mathrm{U}_3(t)\,\mathrm{U}_5(t) - \mathrm{U}_2(t)\,\mathrm{U}_4(t)\,\mathrm{U}_5(t) - \mathrm{U}_2(t)\,\mathrm{U}_4(t)\,\mathrm{U}_4(t) \\
& - 2\,\mathrm{U}_1(t)\,\mathrm{U}_2(t)\,\mathrm{U}_3(t)\,\mathrm{U}_4(t)\,\mathrm{U}_5(t) + \mathrm{U}_1(t)\,\mathrm{U}_2(t)\,\mathrm{U}_4(t)\,\mathrm{U}_5(t) \\
& - \mathrm{U}_2(t)\,\mathrm{U}_3(t)\,\mathrm{U}_4(t)\,\mathrm{U}_5(t) + \mathrm{U}_1(t)\,\mathrm{U}_3(t)\,\mathrm{U}_4(t)\,\mathrm{U}_5(t) \\
& + \mathrm{U}_1(t)\,\mathrm{U}_2(t)\,\mathrm{U}_3(t)\,\mathrm{U}_5(t) + \mathrm{U}_1(t)\,\mathrm{U}_2(t)\,\mathrm{U}_3(t)\,\mathrm{U}_4(t).
\end{aligned}$$

4.3 Fault trees

4.3.1 General concept

Although fault trees are very similar to RBDs, their probabilistic approach is the logical opposite. Instead of modeling components that need to be available in order to keep a SIS running, they outline combinations of unavailable components that would result in overall system unavailability. Basically, FTs are also graphs, defined by a tuple. In this case a three-tuple according to

$$\mathcal{FT}\left(\mathcal{P},\mathcal{R},\mathcal{E}\right),$$

where $\mathcal{P}$ and $\mathcal{R}$ are the graph's vertices, and $\mathcal{E}$ are directed edges. In graph drawings, the direction of the edges is usually not explicitly denoted, as FTs and thus the edges are typically directed from bottom to top. FT vertices can be distinguished into:

- primary failure events, i.e., component failures $\mathcal{P}$ (depicted as circles)
- logical relations over primary events and/or intermediate data from other relations $\mathcal{R}$ (depicted as boxes)

Additional (rounded) boxes are allowed for commenting purposes (see, e.g., fig. 4.5: 'SIS failed'). But as they do not serve any mathematic purpose, they are left out from all subsequent treatments. Valid failure forwarding paths $\mathcal{E}$, i.e., the graph's edges, are

$$\mathcal{E} \subseteq \left((\mathcal{P}\times\mathcal{R})\cup(\mathcal{R}\times\mathcal{R})\right).$$

Only one vertex in the graph is allowed to not have any abducent edge, rendering this vertex the system output:

$$(\exists! r_{out} \in (\mathcal{P}\cup\mathcal{R}))\left((\nexists\,(r_1,r_2)\in\mathcal{E})\,(r_1 = r_{out})\right).$$

From this vertex, the direction of all edges can implicitly be derived. Basic FTs distinguish three subtypes of relations: AND relations (&), OR relations (≥ 1), and NOT (1) relations. They are defined in the same way as in electrical TTL.

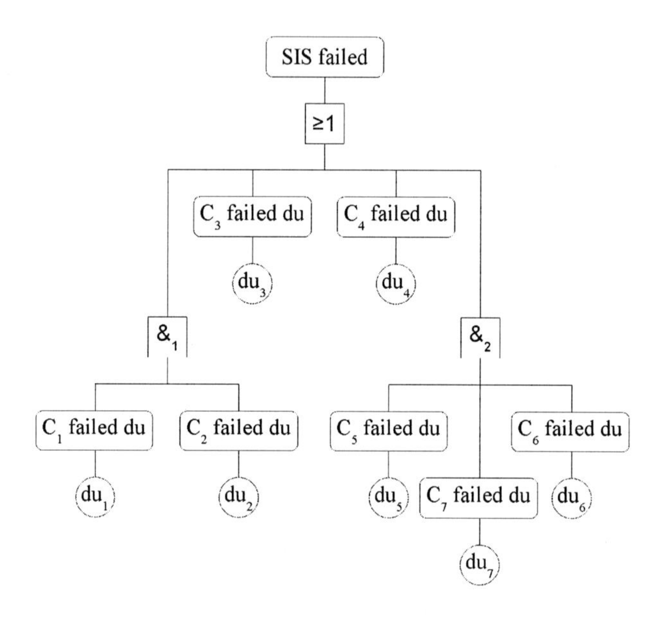

Figure 4.5: Exemplary fault tree

4.3.2 Calculation model

FTs allow for the construction of a system unavailability function $\mathrm{U}_{SIS}(t)$. This is accomplished by decomposing the FT, starting with the system output, and

- replacing relation block outputs with mathematical expressions that depend on the relation type as well as the block inputs
- replacing primary failure event block outputs with the related component unavailability functions

In accordance to eqs. 4.1 and 4.2 for RBDs (p. 44), it is easy to derive unavailability functions representing the outputs of the relation blocks:

$$\mathrm{U}_{AND}(t) = \mathrm{f}(\mathrm{U}_1(t), \ldots, \mathrm{U}_i(t), \ldots, \mathrm{U}_n(t)) = \prod_{i=1}^{n} \mathrm{U}_i(t) \tag{4.3}$$

$$\mathrm{U}_{OR}(t) = \mathrm{f}(\mathrm{U}_1(t), \ldots, \mathrm{U}_i(t), \ldots, \mathrm{U}_n(t)) = 1 - \prod_{i=1}^{n} (1 - \mathrm{U}_i(t)) \tag{4.4}$$

$$\mathrm{U}_{NOT}(t) = \mathrm{f}(\mathrm{U}_1(t)) = 1 - \mathrm{U}_1(t) \tag{4.5}$$

The equations hold for AND and OR blocks with n inputs each, and NOT blocks with one input. For the example outlined in fig. 4.5 this results in

$$\mathrm{U}_{SIS}(t) = 1 - \left[(1 - \mathrm{U}_{\&_1}(t))\,(1 - \mathrm{U}_3(t))\,(1 - \mathrm{U}_4(t))\,(1 - \mathrm{U}_{\&_2}(t))\right],$$

where

$$\mathrm{U}_{\&_1}(t) = \mathrm{U}_1(t)\,\mathrm{U}_2(t),$$
$$\mathrm{U}_{\&_2}(t) = \mathrm{U}_5(t)\,\mathrm{U}_6(t)\,\mathrm{U}_7(t).$$

Comparing eqs. 4.3 and 4.4 with eqs. 4.1 and 4.2 for RBDs, an obvious analogy between both methods becomes evident. It is generally possible to transform every RBD into an adequate FT and vice versa by replacing

- Series structures in RBDs with OR relations in FTs
- Parallel structures in RBDs with AND relations in FTs
- Bridge structures in RBDs with suitable OR-NOT combinations in FTs (reflecting the sum of products principle)

as shown in [B"o06]: p. 117. The FT in fig. 4.5 serves as an example for that since it perfectly matches the RBD in fig. 4.1 (p. 42). At this point it is already obvious that the explosion of the number of combinatorial terms with increasing number of components is a major drawback of the combinatorial approaches. However, effective algorithms are available, that allow for efficient decomposition of the RBDs or FTs. These perform optimized searches for the so-called minimal cutsets, i.e., all minimal subsets of failed components leading to a system failure.

4.3.3 Numerical PFD_{avg}-derivation

The SIS's unavailability function is typically difficult to express in closed mathematical form. Thus, it is in most cases not suitable to construct

$$PFD_{avg} = \frac{1}{T_{obs}} \int_0^{T_{obs}} \mathrm{U}_{SIS}(t)dt. \tag{4.6}$$

Instead, the time axis is discretized and by utilizing fast algorithms, a numerical solution based on the component individual unavailability functions in the form of lookup tables can be obtained. For each discrete time step k, the output of each relation block can be calculated by referring to the matching lookup table values assigned to the primary event boxes and combine them according to eqs. 4.3 to 4.5. That way the SIS's unavailability function $\mathrm{U}_{SIS}(k) = \mathrm{PFD}(k)$ can be obtained. Hence,

$$PFD_{avg} = \frac{\Delta t}{T_{obs}} \sum_{k=0}^{\frac{T_{obs}}{\Delta t}-1} \mathrm{U}_{SIS}(k). \tag{4.7}$$

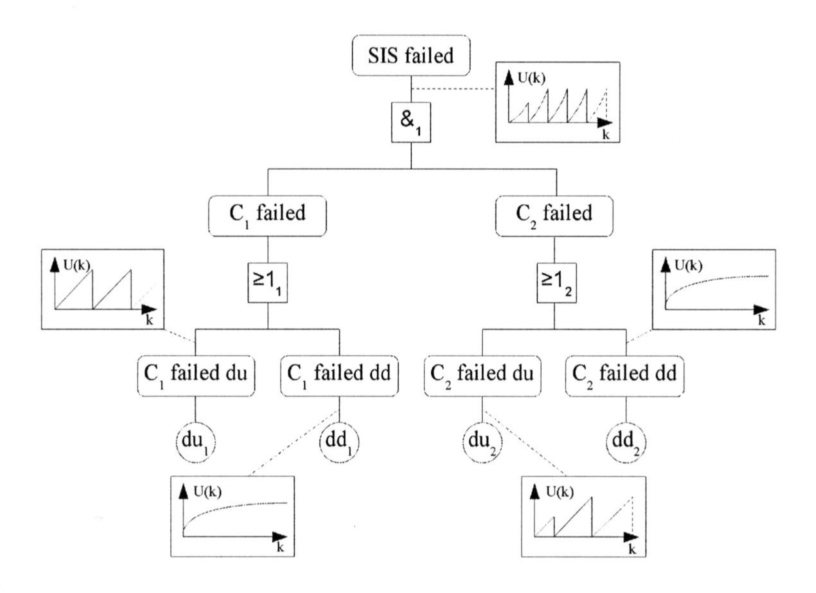

Figure 4.6: Exemplary PFD_{avg}-derivation based on lookup tables

immediately follows. Notice that in the exemplary fig. 4.6 two different RFR cycle types are used. The first one matches the RFR-DRI process according to subsec. 3.5.3 (p. 38). Here, the adhoc knowledge about proof test times is utilized explicitly in the related unavailability functions (instant decrease in unavailability after each $n\frac{T_I}{\Delta t}$). This RFR cycle represents *du* failures. The other RFR process type matches instant repair initiation according to subsec. 3.5.1 (p. 36), representing *dd* failures. Notice that the information about proof tests is not utilized here (missing instant decrease after the proof test procedures). The resulting system unavailability function therefore does not fall back to zero after the proof tests, as the *dd* failures always contribute with a certain amount of unavailability. It is of course possible to extend the *dd* failure's unavailability functions adequately.

4.3.4 Simplified PFD_{avg}-derivation: equidistant proof tests

Sufficiently simple conditions enabling for a closed expression of the system's unavailability function are typically met, if

- only RFR-DRI processes of type B according to subsec. 3.5.3 (p. 38) are relevant for the calculation,

- all components are proof tested at the same equidistant points in time without staggering,
- the observation time equals multiples of the proof test period: $T_{obs} = nT_I$.

If the observation time is large compared to the proof test interval, then the last prerequisite approximatively holds for arbitrary T_{obs} (see also eq. 3.14). Typically, this approximation always holds since the observation time is chosen as the SIS's mission time. Then the system's unavailability $\mathrm{U}_{SIS}(t)$ between proof tests equals its unreliability $\mathrm{F}_{SIS}(t)$, as no maintenance is applied to any component meanwhile. Thus, $\mathrm{U}_{SIS}(t)$ is cyclic with the proof test period T_I, and PFD_{avg} can be derived as the arithmetic mean of the SIS's unreliability function over the first proof test period T_I.

$$\mathrm{PFD}_{avg} = \frac{1}{T_I}\int_0^{T_I} \mathrm{F}_{SIS}(t)dt.$$

This expression is often easy to handle, since only the components' exponential unreliability functions are used. This approach is particularly suitable for smaller (homogeneously instrumented) systems, as $\mathrm{F}_{SIS}(t)$ then condenses into a relatively compact expression. For a simple 1oo1 system according to fig. 4.7 with unreliability $\mathrm{F}_1(t) = 1 - \exp(-\lambda t)$ this leads to

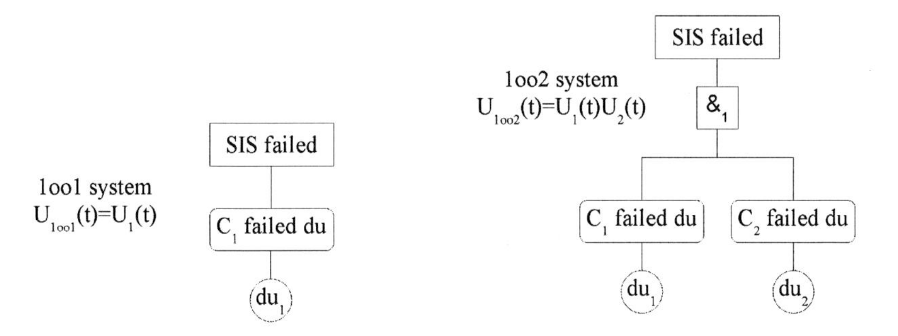

Figure 4.7: Fault tree for a simple 1oo1 system

Figure 4.8: Fault tree for a simple 1oo2 system

$$\begin{aligned}\mathrm{PFD}_{avg,1oo1} &= \frac{1}{T_I}\int_0^{T_I} \mathrm{U}_1(t)dt = \frac{1}{T_I}\int_0^{T_I} \mathrm{F}_1(t)dt \\ &= \frac{1}{T_I}\int_0^{T_I} 1 - \exp(-\lambda t)\, dt \\ &= 1 + \frac{\exp(-\lambda T_I) - 1}{\lambda T_I}\end{aligned}$$

(necessarily identical with eq. 3.14 (p. 39)). Expanding this result into a MacLaurin series, the expected term as derived in subsec. 3.5.3 (p. 38) follows:

$$\mathrm{PFD}_{avg,1oo1} \approx \frac{1}{2}\lambda T_I.$$

Other PFD formulas can be derived accordingly. From the FT in fig. 4.8, the 1oo2 unavailability function $\mathrm{U}_{1oo2}(t)$ can be obtained as

$$\mathrm{U}_{1oo2}(t) = \mathrm{U}_1(t)\,\mathrm{U}_2(t).$$

Thus, the average unavailability over the first proof test interval can be calculated as

$$\begin{aligned}\mathrm{PFD}_{avg,1oo2} &= \frac{1}{T_I}\int_0^{T_I} \mathrm{U}_1(t)\,\mathrm{U}_2(t)dt = \frac{1}{T_I}\int_0^{T_I} \mathrm{F}_1(t)\,\mathrm{F}_2(t)dt \\ &= \frac{1}{T_I}\int_0^{T_I} \left(1-\exp\left(-\lambda t\right)\right)^2 dt.\end{aligned}$$

From here on, the rest of the derivation is straight forward, resulting in

$$\mathrm{PFD}_{avg,1oo2} \approx \frac{1}{3}\lambda^2 T_I^2.$$

It is important to point out again that this closed mathematic expression can only be derived if radical simplifications are applied, e.g., conducting proof tests non-staggered at equidistant points in time while neglecting *dd* failures totally.

Combinatorial approaches as introduced in this section all assume independent unavailability functions and thus independent maintenance teams. The validity of this fact has to be evaluated for concrete scenarios.

4.4 Markov models

4.4.1 General concept

Basically, markov models (MMs) are autonomous stochastic automata. Therefore, they differ from classic automata such as the well known deterministic and finite state machines as used, e.g., for the design of SPLC code. MMs are used to calculate the time-dependent probability for being in any of the various internal states of a SIS. The specific follow-up state to be entered is not determinable and depends on the failure behavior of the underlying components. Hence the name: stochastic automaton. MMs do not support a workflow like the combinatorial approaches, i.e., firstly derive a calculation model representing the logical interdependence of the components, and subsequently retrieve the desired target characteristics. Instead, the basic automaton construction decides upon what type of target characteristic is available. MMs for unreliability calculation differ from MMs for unavailability calculation. However, a conversion of unavailability into unreliability MM is possible by applying simple transformation formulas. Thus, instead of presenting a

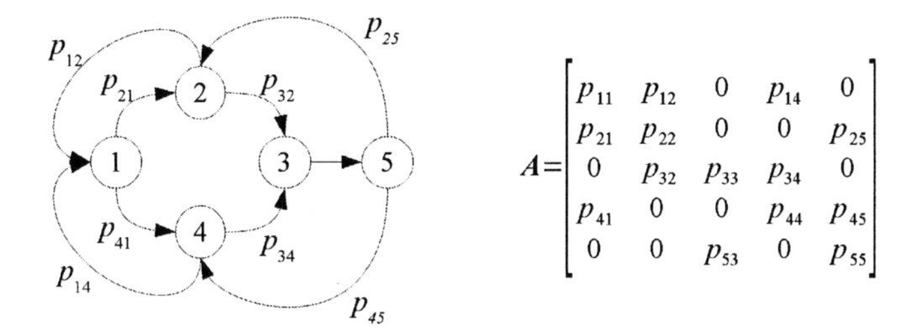

Figure 4.9: Simple exemplary markov model

closed approach, different strategies for setting up unreliability as well as unavailability models are given in the subsequent paragraphs.

For reliability and availability analysis, the most important MM type with respect to this thesis is the *discrete time homogeneous MM.* It is defined as a four-tuple

$$\mathcal{MM}(\mathcal{S}, \boldsymbol{A}, \boldsymbol{p}_0, \boldsymbol{c}^\top).$$

It consists of a set of states $\mathcal{S}$ with $|\mathcal{S}| = Q$, representing each possible system state the SIS may enter. For keeping things simple, it is valid to define the elements in $\mathcal{S}$ as state space vectors $\boldsymbol{s}_i$ according to

$$\begin{aligned} &\mathcal{S} = \{\boldsymbol{s}_1, \ldots, \boldsymbol{s}_Q\} \\ &\text{with } \boldsymbol{s}_i \in \left\{(\boldsymbol{s}[1] \ \ldots \ \boldsymbol{s}[N_{SIS}])^\top \mid (\boldsymbol{s}[1], \ldots, \boldsymbol{s}[N_{SIS}]) \in (\mathcal{CS} \times \mathcal{CS} \times \ldots \times \mathcal{CS})\right\}. \end{aligned}$$

N_{SIS} is the number of components in the SIS. Each component has a distinct internal component state for each system state $\boldsymbol{s}_i \in \mathcal{S}$. $\boldsymbol{s}_i[j] \in \mathcal{CS}$ [1] denotes the state of component j when the system is in state i. The possible component states in $\mathcal{CS}$ vary depending on the desired modeling depth. Typically, $\{OK, DU, DD\} \subseteq \mathcal{CS}$ is defined.

An initial probability distribution vector $\boldsymbol{p}_0$ of size $Q \times 1$ with $\sum_i \boldsymbol{p}_0[i] = 1$ is used to describe the initial condition of the SIS, i.e., at time step $k = 0$. Usually the initial state is deterministically known and thus one $\boldsymbol{p}_0[i] = 1$ and all $\boldsymbol{p}_0[j \neq i] = 0$. From the initial state the system will traverse through its state space. The specific trace is not determinable due to the stochastic character of the automaton. However, the SIS's failure dynamics are represented in the $Q \times Q$ transition matrix $\boldsymbol{A}$. The elements $\boldsymbol{A}[i][j]$ encode the probability for the system to be in state i at time step $k+1$ given the system in state j at time step k. These probabilities are constant over time, hence the term 'homogeneous'. An element $\boldsymbol{A}[i][j] \neq 0$ can also be interpreted as an edge (j, i) in the related MM graph, where every state $s_i \in \mathcal{S}$ is represented by a vertex (see fig. 4.9). Typically, all recursive

[1]The component state set $\mathcal{CS}$ will be defined in a more formal way in subsec. 5.2.1 (p. 72)

edges, i.e., transitions from any state back to the same state, are not explicitly drawn in the graph. Their probabilities can be derived as

$$\boldsymbol{A}[i][i] = 1 - \sum_{j \neq i} \boldsymbol{A}[j][i].$$

This means that after each discrete time step of length Δt the system must perform a transition which might either be back to the currently occupied state (via recursion) or to any other possible subsequent state.

In order to derive arbitrary reliability or availability parameters, the probabilities of all states contributing to the desired target characteristic have to be selected. This can be achieved by utilizing a $1 \times Q$ selection vector $\boldsymbol{c}$ with

$$\boldsymbol{c}[i] = \begin{cases} 1 & \text{if } \boldsymbol{s}_i \text{ counts for the desired parameter} \\ 0 & \text{else.} \end{cases} \tag{4.8}$$

The probability for being in any of its Q states at any point in time is expressed by the probability distribution vector $\boldsymbol{p}(k)$ according to $\boldsymbol{p}[i](k) = \mathrm{P}(\text{System state} = s_i \in \mathcal{S})(k)$. Then the desired time-dependent reliability/availability characteristic $\mathrm{f}(k)$ can be derived as

$$\mathrm{f}(k) = \boldsymbol{c}^\top \boldsymbol{p}(k).$$

The evolution of the system over time can be expressed by the recursive Chapman-Kolmogorov equation

$$\boldsymbol{p}(k+1) = \boldsymbol{A}\boldsymbol{p}(k)$$

which can be transformed into the absolute equation:

$$\boldsymbol{p}(k) = \boldsymbol{A}^k \boldsymbol{p}_0.$$

With that and utilizing the selection vector (eq. 4.8) again for picking relevant states,

$$\mathrm{f}(k) = \boldsymbol{c}^\top \boldsymbol{p}(k) = \boldsymbol{c}^\top \left(\boldsymbol{A}^k \boldsymbol{p}_0\right)$$

immediately follows. Discrete time homogeneous MMs as introduced here represent exponentially distributed lifetime processes. This is possible, as these processes are the only ones featuring a constant probability of failing during an infinitesimal small time period of time Δt independent from the absolute time t. Of great importance is the choice of the step size Δt as it directly impacts on the quality of the approximation of the underlying real probabilistic process. As repair rates for most failures are around $0.125\,\mathrm{h}^{-1}$, $\Delta t = 1\,\mathrm{h}$ is often suggested. Extended investigations are conducted in, e.g., [FF10], and [Rau04].

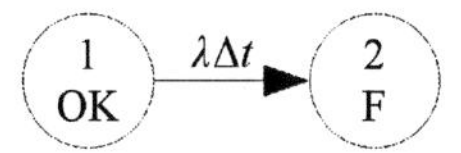

Figure 4.10: MM for unreliability calculation

4.4.2 Unreliability calculation

Unreliability models based on the exponential distribution can be constructed by utilizing absorbing MMs. Their graphs contain states that cannot be left, i.e., $\boldsymbol{A}[i][i] = 1$ for all absorbing states $\boldsymbol{s}_i$. For $k \rightarrow \infty$ the sum of the state probabilities of all absorbing states approaches 1. As can be seen from fig. 4.10, a single component's unreliability function can be modeled as a two state MM. The component starts with $\boldsymbol{p}_0 = (1\,0)^\top$ in state $\boldsymbol{s}_1 = (OK)^\top$, representing the functional state of the component. Afterwards it may fail and enter the absorbing state $\boldsymbol{s}_2 = (F)^\top$. Originating from a simple balance equation, the probability for being in the second state at time step $k+1$ can be expressed as

$$p_2(k+1) = p_2(k) + \lambda\Delta t \cdot p_1(k). \tag{4.9}$$

Here, $\lambda\Delta t$ is the probability for the component to leave the good state between the k-th and the $(k+1)$-th time step. From this term it is obvious that the underlying failure process must be exponentially distributed, as $\mathrm{P}\left((k\Delta t < TTF \leq (k+1)\Delta t) \mid (TTF > k\Delta t)\right) = \lambda\Delta t$ only holds for the exponential distribution. Equation 4.9 can be transformed according to

$$\frac{p_2(k+1) - p_2(k)}{\Delta t} = \lambda \cdot p_1(k).$$

This expression can be interpreted as an Euler forward discretization of a time continuous term. With $p_1(t) = 1 - p_2(t)$

$$\frac{dp_2(t)}{dt} = \lambda\left(1 - p_2(t)\right).$$

follows. The solution for this equation is the well known unreliability function $p_2(t) = 1 - \exp(-\lambda t) = \mathrm{F}(t)$. $\lim_{t\rightarrow\infty} p_2(t) = 1$, as expected for the absorbing MM. This method of explaining the behavior of discrete time MMs by solving the related time continuous ordinary differential equation (ODE) will be applied again in the next subsections. The time continuous expression allows for direct comparisons with the basic concepts of unavailability and unreliability calculation as introduced in subsecs. 3.3 and 3.5.

4.4.3 Unavailability calculation: RFR-IRI process

The MM as shown in fig. 4.11 is used to model the unavailability function if an RFR cycles with instant repair initiation as introduced in subsec. 3.5.1 (p. 36) is applied. According

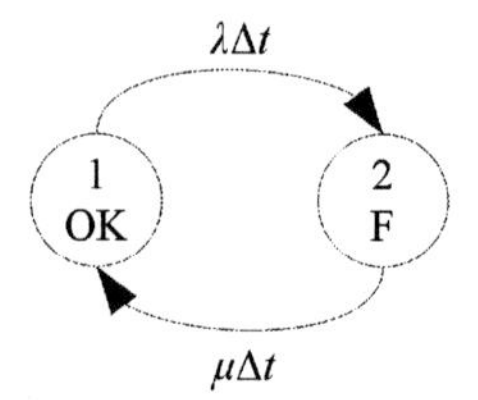

Figure 4.11: MM for RFR-IRI unavailability calculation

to the unreliability calculation model, a differential equation for the second state can be derived as

$$\begin{aligned} p_2(k+1) &= p_2(k) + \lambda\Delta t \cdot p_1(k) - \mu\Delta t \cdot p_2(k) \\ \frac{p_2(k+1) - p_2(k)}{\Delta t} &= \lambda - p_2(k)\left[\lambda + \mu\right] \end{aligned}$$

and with that

$$\frac{dp_2(t)}{dt} = \lambda - p_2(t)\left[\lambda + \mu\right].$$

The solution to this ODE is

$$p_2(t) = \mathrm{U}(t) = \frac{\lambda}{\lambda + \mu}\left[1 - \exp\left(-\left(\lambda + \mu\right)t\right)\right],$$

which perfectly matches eq. 3.10 (p. 36).

4.4.4 Unavailability calculation: RFR-DRI process - type A

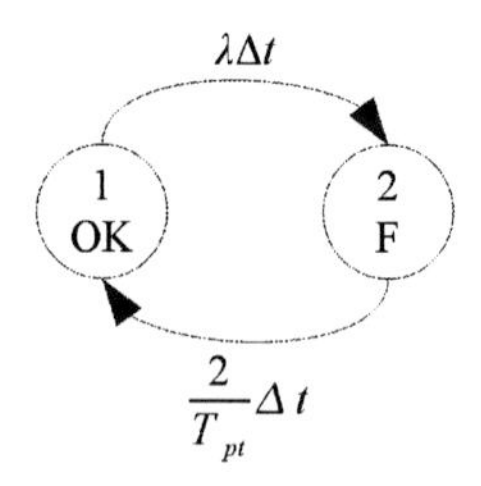

Figure 4.12: MM for type A RFR-DRI unavailability calculation

RFR cycles with delayed repair initiation are difficult to apply to standard MMs. As shown in the previous sections, MMs reproduce exponentially distributed random processes only. Therefore, it is challenging to integrate a deterministic process such as a regularly conducted proof test into the models. One commonly utilized method is to fall back to the type A RFR-DRI approach as introduced in subsec. 3.5.2 (p. 36). The original RFR-DRI process is approximated by losing relation to absolute time and parameterizing an RFR-IRI process with appropriate expected value. This approach is nowadays seldomdly utilized, as the multiphase markov approach (see subsequent subsec. 4.4.5) is much better suited. However, the said method is applied in, e.g., [Hil07a]. The MM according to fig. 4.12 delivers the ODE

$$\frac{dp_2(t)}{dt} = \lambda - p_2(t)\left[\lambda + \frac{2}{T_I}\right],$$

with solution

$$p_2(t) = \mathrm{U}(t) = \frac{\lambda}{\lambda + \frac{2}{T_I}}\left[1 - \exp\left(-\left(\lambda + \frac{2}{T_I}\right)t\right)\right].$$

This equals exactly the unavailability over time according to eq. 3.13 (p. 38) and can easily be transformed into an average unavailability.

4.4.5 Unavailability calculation: RFR-DRI process - type B and C

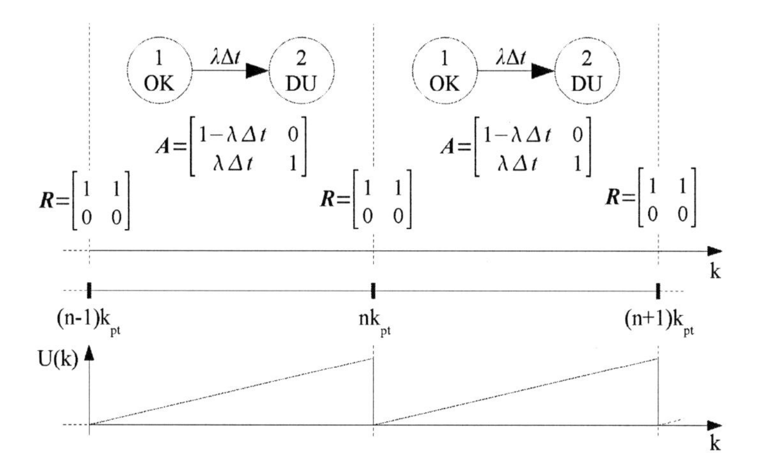

Figure 4.13: multiphase markov model

Another approach of modeling time driven repair with MMs is the multiphase extension. The basic idea is to split the discrete time axis into phases of arbitrary length.

Between phase transitions the system is described by standard MMs. These may be identical for each phase or differ in size and structure. The only requirement is that the initial condition for the upcoming phase can be determined explicitly. Usually it is derived algorithmically from the last probability distribution of the preceding phase. Multiphase markov models are a well established method to represent periodically proof tested systems. Various publications deal with the analysis and the application of this mathematical approach, such as, e.g., [Buk01], [Buk05], [GC05], [Inn08], and [GLS08a].

The example in fig. 4.13 splits at multiples of the discrete proof test interval $k_{pt} = \frac{T_I}{\Delta t}$. During each phase the system is described by a standard two-state unreliability MM. The probability distribution vector $\boldsymbol{p}(k = nk_{pt} - 1)$, i.e., right before each phase transition, is used to calculate the initial distribution for the MM of the subsequent phase. This is accomplished by utilizing a probability shifting matrix $\boldsymbol{R}$ of dimension $Q \times Q$ (with $Q = |\mathcal{S}|$) according to

$$\boldsymbol{p}(nk_{pt}) = \boldsymbol{R}\boldsymbol{A}\boldsymbol{p}(nk_{pt} - 1) \text{ with } \boldsymbol{R} = \begin{bmatrix} 1 & 1 \\ 0 & 0 \end{bmatrix},$$

which performs the transition from phase $n-1$ to phase n (see also, e.g., [GC05]). $\boldsymbol{R}$ shifts probability from the second (failed) state back to the first (functional) state:

$$\begin{aligned} \boldsymbol{p}[1](nk_{pt}) &= (\boldsymbol{A}[1,1] + \boldsymbol{A}[2,2])\boldsymbol{p}[1](nk_{pt} - 1) + (\boldsymbol{A}[1,2] + \boldsymbol{A}[2,1])\boldsymbol{p}[2](nk_{pt} - 1) \\ \boldsymbol{p}[2](nk_{pt}) &= 0. \end{aligned}$$

According to this notation, the phase transition consists of a last system evolution via Chapman-Kolmogorov recursion as well as the shifting process itself. In order to retrieve valid probability distribution vectors,

$$\forall j \left(\sum_i \boldsymbol{R}[i][j] = 1 \right)$$

must be guaranteed. Typically, $\boldsymbol{R}[i][i] = 1$, if the probability that has cumulated in state i must remain after the phase transition. $\boldsymbol{R}[i][j] = 1$, if the cumulated probability of state j must be shifted to state i. The whole procedure can be described in closed mathematical form, introducing a new definition:

Definition 4.1 (Multiphase markov model). *A multiphase markov model (MPMM) suitable for reproducing an RFR-DRI process of type B, can be defined as*

$$\mathcal{MPMM}\left(\mathcal{S}, \boldsymbol{A}, \mathbf{p}_0, \mathbf{c}^\top, \boldsymbol{R}, k_{pt}\right).$$

The discrete time unavailability $\mathrm{U}(k)$ *of this MPMM with state space* $\mathcal{S}$*, periodical proof tests each* nk_{pt}*, a transition matrix* $\boldsymbol{A}$*, an initial probability distribution* $\mathbf{p}_0$*, a selection vector* $\mathbf{c}^\top$*, and a probability shifting matrix* $\boldsymbol{R}$*, can be expressed recursively as*

$$\mathrm{U}(k) = \mathbf{c}^\top \boldsymbol{p}(k), \boldsymbol{p}(0) = \mathbf{p}_0$$

$$\boldsymbol{p}(k+1) = \begin{cases} \boldsymbol{R}\boldsymbol{A}\boldsymbol{p}(k) & \textit{if } ((k+1) \bmod k_{pt}) = 0, \\ \boldsymbol{A}\boldsymbol{p}(k) & \textit{else.} \end{cases}$$

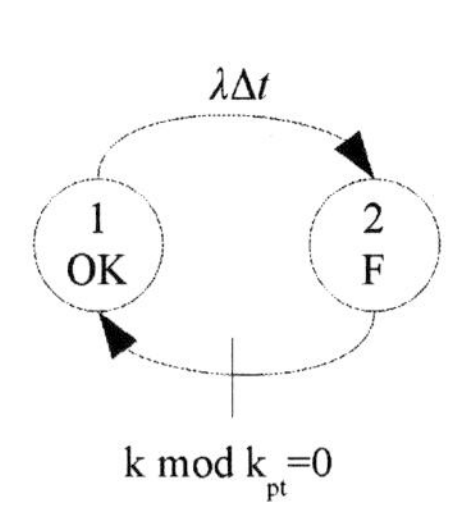

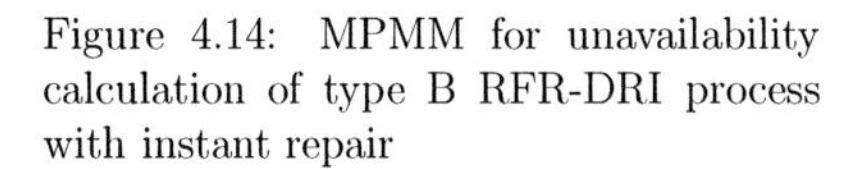
Figure 4.14: MPMM for unavailability calculation of type B RFR-DRI process with instant repair

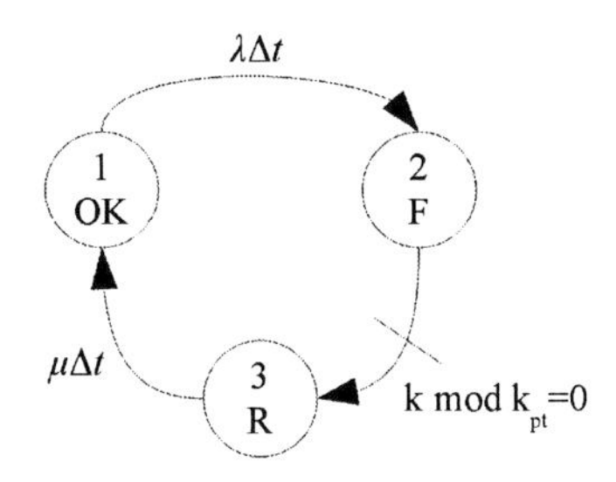

Figure 4.15: MPMM for unavailability calculation of type C RFR-DRI process with exponential repair time

This MPMM perfectly reproduces the unavailability of an RFR-DRI process of type B according to subsec. 3.5.3 (p. 38).

Figure 4.14 and 4.15 introduce a new notation for outlining the multiphase character of the MM with periodic phase transitions. The transitions marked with a crossing line and the label $k \bmod k_{pt} = 0$ define the cycle length of each phase, staggering (for $k \bmod k_{pt} \neq 0$), and simultaneously denote the target state of the shifting process. It is therefore possible to obtain the entire calculation model related to the MPMM depicted in fig. 4.14.

Figure 4.15 shows an implementation of the standard type C RFR-DRI process according to subsec. 3.5.4 (p. 40). An additional repair state 3 is inserted. At phase transitions, probability is shifted from the failure state 2 to the repair state 3 from where a standard FR process with repair rate μ leads back to the repaired state 1. Notice that a pure closed mathematical expression for this process is not trivial which clearly outlines the powerful modeling capabilities of markovian approaches.

The resulting unavailability can thus be derived with a selection vector according to $c^{\top} = (0\,1)$ for the MPMM in fig. 4.14 and $c^{\top} = (0\,1\,1)$ for fig. 4.15, respectively. The latter one assumes that the component is not functional while under repair.

4.4.6 Multi-component systems

In larger systems, the two important RFR processes are still easily applicable. Consider a system according to fig. 4.16. The MM represents the two superimposed unavailability functions of the modeled components. The related ODEs are

$$\frac{dp_1(t)}{dt} = \mu_1 p_2(t) + \mu_2 p_3(t) - [\lambda_1 + \lambda_2]\, p_1(t),$$
$$\frac{dp_2(t)}{dt} = \lambda_1 p_1(t) + \mu_2 p_4(t) - [\lambda_2 + \mu_1]\, p_2(t),$$
$$\frac{dp_3(t)}{dt} = \lambda_2 p_1(t) + \mu_1 p_4(t) - [\lambda_1 + \mu_2]\, p_3(t),$$

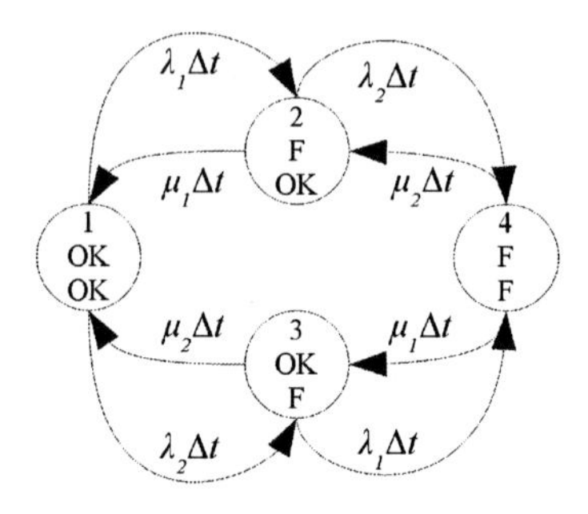

Figure 4.16: MM representing a two component SIS

$$\frac{dp_4(t)}{dt} = \lambda_2 p_2(t) + \lambda_1 p_3(t) - [\mu_1 + \mu_2]\, p_4(t).$$

As, e.g., the first component has failed in second and fourth state of the MM, its unavailability can be constructed according to

$$\mathrm{U}_1(t) = p_2(t) + p_4(t).$$

This new variable can be obtained from the ODE system:

$$\begin{aligned} \frac{d\,(p_2(t) + p_4(t))}{dt} &= \lambda_1 \left(\underbrace{p_1(t) + p_3(t)}_{=1-(p_2(t)+p_4(t))} \right) - \mu_1\,(p_2(t) + p_4(t)) \\ \frac{d\,\mathrm{U}_1(t)}{dt} &= \lambda_1\,(1 - \mathrm{U}_1(t)) - \mu_1\,\mathrm{U}_1(t) \\ &= \lambda_1 - \mathrm{U}_1(t)\,[\mu_1 + \lambda_1]\,. \end{aligned}$$

The solution is

$$\mathrm{U}_1(t) = \frac{\lambda_1}{\lambda_1 + \mu_1} \exp\left[-\left(\lambda_1 + \mu_1\right) t\right].$$

The unavailability of the first component is thus totally independent from the RFR process of the second component, i.e., no interdependencies are present. It is therefore easy to construct arbitrary large MM representing independent component RFR cycles.

4.4.7 Repair team models

The simple the denoted example in the previous subsec. 4.4.6 might appear, the fundamental is the assumption it is based on: *total* independence of the modeled components.

On the one hand, independence is easy to declare for the RF processes. On the other hand, the FR processes, i.e., the restoration processes, cannot automatically be considered independent from each other. The MM as depicted in fig. 4.16 (p. 58) assumes two repair teams, one for each represented component. In reality, only a single repair team is usually available. This leads to the question of how to design MMs that represent a different repair strategy, e.g., a system with a single repair team. This question has been discussed very controversely in, e.g., [Buk05], [Gul03], and [May09]. Suggestions can be found in [PP98], [GC05], [GY08], and even in the most recent edition of IEC 61508 [IEC09] (using a priorized repair strategy). However, the extension to a system of R repair teams for N components seems currently to be not definitely solved. It is therefore valid to stick to the models assuming N repair teams for N components as favored by, e.g., IEC 61165 [DIN04] (a standard about the application of the markov modeling approach referred to by the old and the new edition of IEC 61508), and [Buk05]. Notice that the simplified formulas as derived for IEC 61508 are based on a N repair team for N components model.

4.4.8 Unavailability calculation: competing failure types

Failure type distribution approach

Typically, components have multiple failure types (see R8, as well as A2 to A7). The general way of thinking is assuming an overall failure rate λ_{FTD} for a component with R different failure types. The device fails after its time to failure

$$\mathrm{P}(TTF \leq t) = \mathrm{F}(t) = 1 - exp\left(-\lambda_{FTD} t\right).$$

With the occurrence of the failure, one of the R failure states is entered. The probability p_i for each state S_i to be entered is denoted by the failure type distribution which can be obtained, e.g., via a failure mode and effects analysis (FMEA) of the considered component. As always one of the failure states must be entered after failure occurrence, $\sum_R p_i = 1$ holds.

Example 4.1 (Exemplary component). A component with failure rate $\lambda_{FTD} = 1000\,\mathrm{FIT}$ [2] and a failure mode distribution according to

$$dd: 40\%, du: 25\%, dn: 5\%, a: 30\%,$$

has a probability of, e.g., $p_{du} = 0.25$ that an arbitrary component failure is a du failure.

Markovian approach

This principle is implemented differently in MMs which is seldomly recapitulated in literature. Here, R different failure processes are superimposed (see fig. 4.17). Each with its

[2] 1 FIT equals one failure per 10^9 hours. In process industry, FITs are typically used for expressing failure rates.

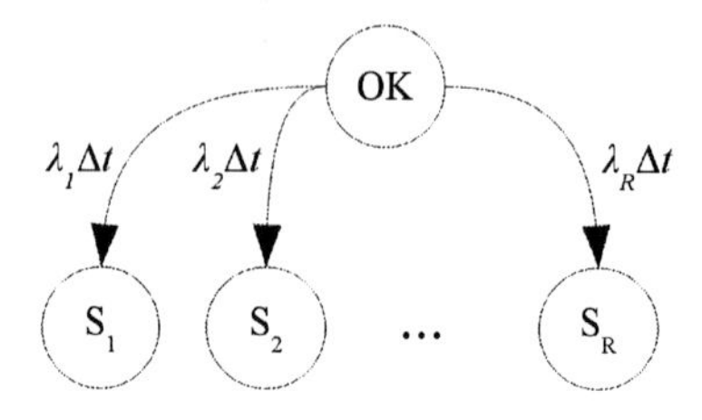

Figure 4.17: MM depicting multiple failure types

own TTF_i with related failure rate λ_i. Obviously, the component leaves the OK state after the shortest TTF_i of all competing failure processes. It can be shown that the random variable

$$TTF' = \min(TTF_1, TTF_2, \ldots, TTF_R),$$

describing the 'winning' failure process, is exponentially distributed with

$$\mathrm{P}(TTF' \leq t) = \mathrm{F}'(t) = 1 - exp\left(-\sum_R \lambda_i t\right).$$

(see sec. 10.3 (p. 169) for an extended derivation). If now

$$\sum_R \lambda_i = \lambda_{FTD},$$

where λ_{FTD} is the overall failure rate from the failure distribution approach, then it becomes clear, that the TTF for both variants is the same. Moreover, if p_i from the failure distribution approach is chosen according to

$$p_i = \frac{\lambda_i}{\lambda_{FTD}},$$

it can be shown that

$$\mathrm{P}(TTF_i = \min(TTF_1, TTF_2, \ldots, TTF_R)) = p_i.$$

This means that the 'winning' failure process which dictates the component's failure type in the MM is entered with a probability equal to the related failure mode fraction in the failure distribution approach. Therefore, both approaches behave perfectly identical.

4.5 Monte carlo simulation

A completely different approach towards unavailability calculation is the monte carlo (MC) method. It is a numerical approach, where random experiments are conducted with either a system model or the real process [Erm75], [HT96]. In contrast to the approaches mentioned before, the MC methods does not deliver a closed mathematical term expressing the target characteristic. Instead, it delivers parameter estimators based on sample data which have to be extracted from the random experiments. Treatable problems can be subdivided into two categories:

- deterministic problems, e.g., solving integrals, sums, ODEs
- stochastical problems, e.g., queuing problems, quality, and reliability assessment

In all cases a correct model of the relevant process is crucial. Especially in the second case, the result's accuracy depends heavily on whether all relevant impacting processes have been considered and modeled correctly. It is, e.g., pointless to describe the RF process of a component which is subject to wear-out effects with an exponential CDF. The MC simulation results will reflect the correct overall system behavior with arbitrary accuracy - but only for the (incorrectly modeled) component with exponential failure behavior.

As the MC approach is based on statistical methods, the desired accuracy of the result depends on the number of repetitions of the underlying random experiment, i.e., the sample size. It is of course possible to retrieve sample data from real random experiments by, e.g., implementing the considered SIS multiple times, and observe them over a long time. Unfortunately, the relevant characteristics (PFD_{avg} etc.) require unrealistically large observation periods or, alternatively, an unrealistically large number of implementations. Typically, the random experiments from which the required sample data is deduced, are therefore conducted with computers *simulating* the real experiments. In software simulations huge amounts of experiments can be executed in short time. Various algorithms exist that allow for further acceleration of the simulations.

4.5.1 Advantages and drawbacks

Computer-based simulations have one major drawback: they are not real random experiments. Up to nowadays, computers cannot generate real random numbers. Instead, uncountable methods for generating pseudo random numbers are available. Depending on the requirements of the problem, the one or the other generator algorithm suits better. One very important criterion is the size of the period of the pseudo random number generator (PRNG). This parameter denotes the amount of random numbers that can be generated without periodicity. The required amount of random numbers in a series of random experiments should be lower than the period size. Otherwise, there exists a potential danger of conducting identical random experiments multiple times.

Along with the constraints introduced by PRNGs, one important advantage needs to be mentioned: it is easy to construct arbitrary distributed random variables from the (in most cases) uniformly distributed random numbers from PRNGs. This is accomplished with the inversion method:

Be F an arbitrary CDF and U a uniformly distributed random variable in $[0, 1)$, then

$$X = \mathrm{F}^{-1}(U)$$

is a random variable with CDF F. Exponentially distributed random lifetimes TTF with failure rate λ can therefore be constructed via

$$TTF = -\frac{1}{\lambda}\ln(1-U).$$

4.5.2 Estimation of the desired characteristic

From the central limit theorem it can be deduced, that any complex simulated random variable can be considered gaussian distributed if the number of random experiments is large. It is therefore easy to derive some important statistical characteristics. If the random experiment has been conducted N times, the N outcomes of the desired random variable can be combined into the sample data vector $X = (x_1\, x_2\, \ldots\, x_N)^{\mathsf{T}}$. The average

$$\overline{X} = \frac{\sum_{i=1}^{N} x_i}{N}, \tag{4.10}$$

is a point estimator for the expected value μ of the unknown gaussian distribution. Its unknown variance σ^2 can be estimated according to

$$S^2 = \frac{\sum_{i=1}^{N}\left(x_i - \overline{X}\right)}{N-1}. \tag{4.11}$$

Finally, a confidence interval with niveau $1-\alpha$ for the real parameter $\hat{X}$ can be constructed with

$$\left[\overline{X} - \Phi^{-1}\left(1-\frac{\alpha}{2}\right)\frac{S}{\sqrt{N}}, \overline{X} + \Phi^{-1}\left(1-\frac{\alpha}{2}\right)\frac{S}{\sqrt{N}}\right],$$

where $\Phi^{-1}(\cdot)$ is the quantile of the gaussian distribution. This reads as follows: From an arbitrary sample data vector X of size N, a confidence interval with niveau $1-\alpha$ can be constructed. This interval contains the real value $\hat{X}$ with a probability of $1-\alpha$. However, it is to verify that the sample data is indeed gaussian distributed. This can be accomplished, e.g., with the well-known Kolmogorov-Smirnov test [Har09].

4.5.3 IEC 61508 2nd. ed. petri nets

With respect to this thesis, Petri Net (PN) extensions can be identified as the most important simulation models. Petri Nets have been used in various fields of research (e.g., discrete event systems [CL08]) since 1962 [Pet62]. Uncountable derivates of the basic PN are commonly used worldwide, differing in representable dynamic and static effects. An ordinary PN is a bipartite graph with places and transitions as vertices, interconnected by edges. The most common extension are the 'tokens' which are used to mark places, i.e., rendering them occupied. PNs become very useful if concurrencies have to be modeled. Here, they are superior to standard automata as they allow for smaller, more compact models. This is accomplished by abandoning the way of thinking that one

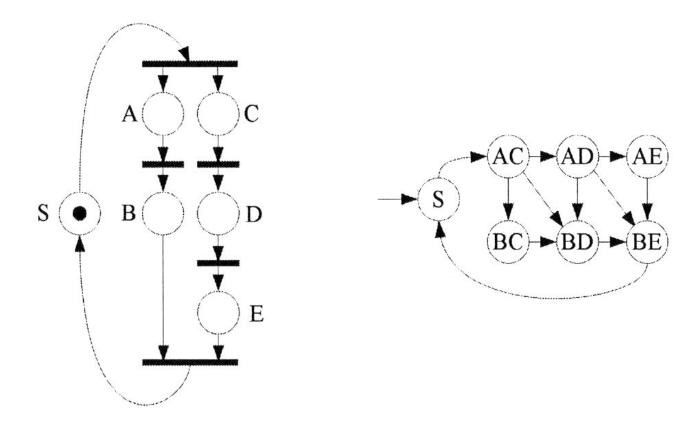

Figure 4.18: Petri net vs. Moore automaton

state of the automaton equals one physical system state. In PNs, the physical system's state is represented by the totality of all marked places which may be more than just one, as the number of tokens in the graph is not generally limited.

Consider fig. 4.18. Here, the possible behavior of the PN on the left is mirrored by a standard Moore automaton. The number of required automaton states is higher than the number of places in the original PN. This problem increases exponentially with the amount of concurrency modeled in the PN. Furthermore, the visual complexity of the PN is significantly lower than the automaton, since the concurrency is apparently visible. The high number of possible state transitions renders the Moore automaton confusing and makes it difficult to interpret its behavior. A simple PN extension, allowing for deterministically as well as stochastically timed component state transitions, is authorized by the new edition of IEC 61508 [IEC09] and appears to be a derivation of generalized stochastic petri nets (GSPN, [MBC84]).

The basic idea is to provide an individual PN for each component of a SIS as shown in fig. 4.19. Additional PNs could be used to model further components or maintenance teams. Specific synchronization as well as timing methods allow for the simulation of real component's behavior and interaction. The PN's transitions are parameterized according to

$$\mathcal{TRANS}\,(\mathcal{PR}, \mathcal{AS}, \delta)\,,$$

where $\mathcal{PR}$ is a set of predicates (denoted as '?' in graphs), $\mathcal{AS}$ is a set of assertions (denoted as '!' in graphs), and δ is a delay specification. The sets might be null sets. Predicates are arbitrary mathematical terms assigned to transitions that have to be evaluated at simulation time in order to activate the related transition. Assertions are conducted immediately after firing the transition they are assigned to. They can be utilized to perform arbitrary variable assignments. As any PN instance may refer to any variable in

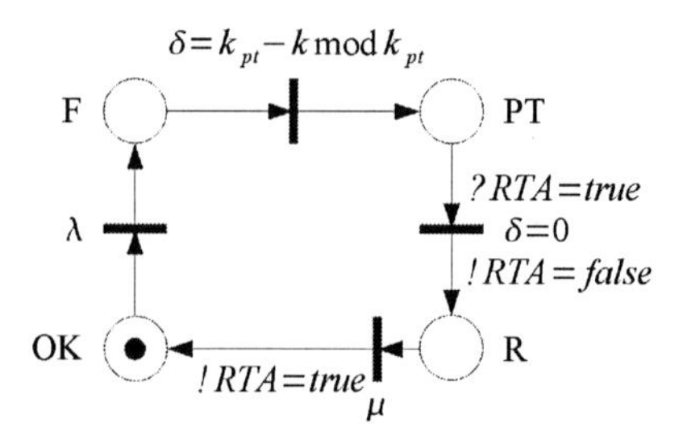

Figure 4.19: PN reproducing an RFR-DRI process

their predicates, a synchronization method arises. Simulation scheduling is enabled by introducing stochastic (denoted as the related exponential CDF's failure rate λ or μ) as well as deterministic delays (denoted as an equation for δ), which have to be evaluated at simulation runtime. The delays express the time a transition has at least to be activated before firing is possible. A transition t_i becomes activated, if

1. one token is in each of its pre places
2. all assigned predicates evaluate as 'true'
3. all post places are empty

The transition fires, if it is activated continuously for a period $\text{eval}\,(\delta_i)$. The firing process includes

1. removing one token from each of the pre places
2. updating all assigned assertions
3. adding one token to each of the post places

Firing transitions are the PN's dynamics generators. The dynamics for the PN depicted in fig. 4.19 will be explained in detail in subsec. 4.5.5. Retrieving safety characteristics from the PN is easy: in a first step the relevant places with respect to the desired characteristic are selected (similar to the concept of selection vectors for MMs). For the PN in fig. 4.19 this includes the states 'failed' (F), 'proof tested' (PT) and 'under repair' (R), if the PFD is to be considered. In the second step, the PN is simulated N times by using appropriate random number generators to evaluate the delays δ. During each experiment, the time of the token spent in PFD-relevant states (downtime DT) is accumulated. In a third step, the specific PFD for each of the N experiments can be calculated according to

$$PFD_i = \frac{\mathrm{DT}_i}{SimulatedTime}$$

(compare to eq. 3.7 (p. 35)). Arranging the N outcomes as

$$X_{PFD} = (PFD_1 \cdots PFD_N)^\top \tag{4.12}$$

finally delivers the sample vector to be treated further with means of stochastical analysis as described above.

4.5.4 Simple RFR-IRI process

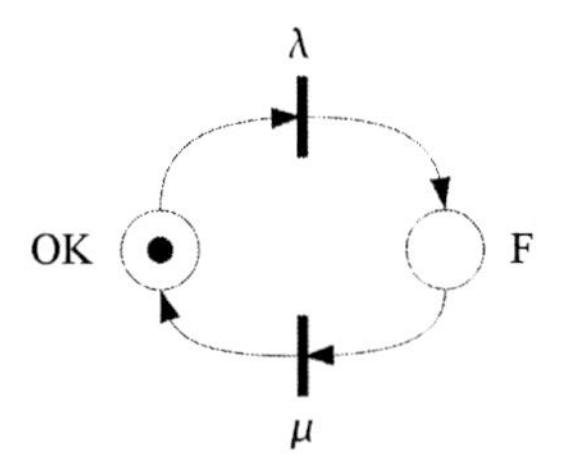

Figure 4.20: PN-model reproducing an RFR-IRI process

The RFR-IRI as well as the three types of RFR-DRI processes as introduced in sec. 3.5 (p. 34) can easily be reproduced with the described PNs. Figure 4.20 depicts the RFR-IRI cycle. Notice that the PN is totally autonomous, as no communication based on shared variables is performed via predicates and assertions. It therefore brings along, e.g., its individual maintenance team.

4.5.5 Simple RFR-DRI processes

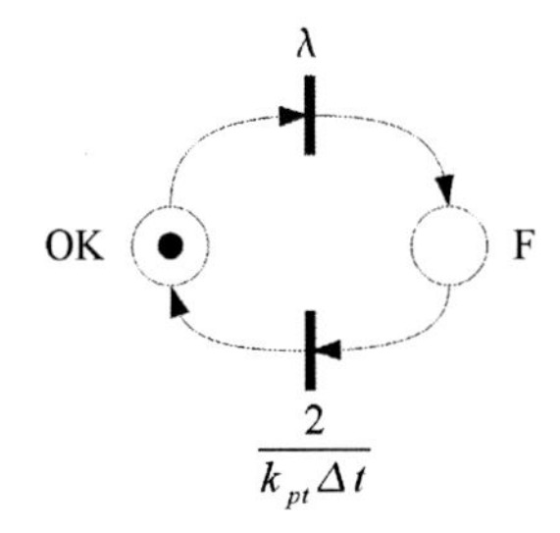

Figure 4.21: PN-model reproducing an RFR-DRI type A process

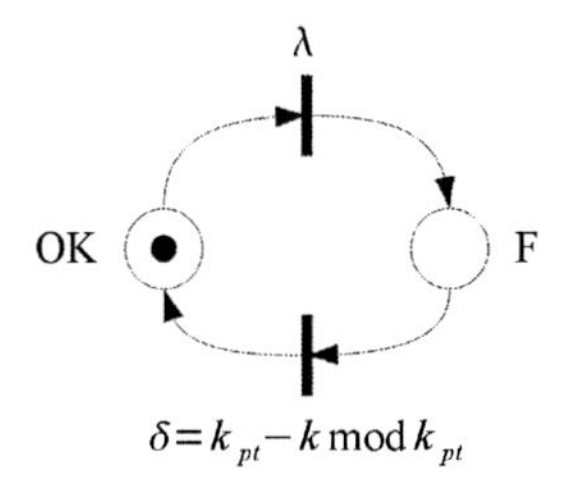

Figure 4.22: PN-model reproducing an RFR-DRI type B process

Figs. 4.21 and 4.22 denote the RFR-DRI processes for both type A and B according to subsec. 3.5.2 (p. 36) and subsec. 3.5.3 (p. 38). The implementation is straight forward and should be self-explanatory. $k_{pt} - (k \bmod k_{pt})$ in fig. 4.22 denotes a semi-deterministic delay, as it specifies the remaining time from the transition activation time k until the upcoming proof test at nk_{pt}.

The PN simulation approach truly shines, when interdependencies of component PNs via maintenance teams are to be modeled. These complex relations are difficult to model with combinatorial approaches or markov theory. A very simple approach that does not rely on a separate PN for maintenance teams is depicted in fig. 4.19 from p. 64. The underlying type C RFR-DRI process according to subsec. 3.5.4 (p. 40) provides extended details that are not available in most approaches: the RF process is specified by the failure rate λ which is equal to the standard approaches so far. The subsequent RF process is modeled rather complex. At first, the component remains in its undetected state until the upcoming proof test is conducted. Afterwards, the repair can only be initiated, if a repair team is available ($RTA = true$ - repair team available). If that is the case, the global variable RTA is set to false while the repair is conducted with exponential CDF (failure rate μ). While $RTA = false$, no further instance of the component PN may initiate a repair process. After finishing the restoration process, RTA is set back to $true$. If this PN is instantiated multiple times, they interact via the global RTA variable which serves the purpose of an interlock semaphor.

4.6 Evaluation towards applicability

4.6.1 Review of requirements

In subsec. 2.4 (p. 24), a number of requirements and assumptions have been introduced. These can now be reviewed in order to evaluate all provided mathematical approaches towards their applicability in the generic model generation method to be developed in this work.

Accessibility

Combinatorial approaches allow for easy access to unavailability modeling, since they immediately provide visual insight into the mutual dependencies between components. Bottlenecks in reliability can easily be identified, the whole layout of the related graphs is close to real component layouts (such as wiring diagrams compared with RBDs). State based approaches on the other hand hide a large amount of structural information. MMs for instance condense component dependencies, i.e., redundancies or sequentially arrangement information, into the selection vectors. This makes it very difficult to 'read' state based approaches. PNs as utilized for the MC approach provide very good insights into component internal dynamics but do also lack easy access to the devices' interactivity.

Component based modeling

R1 (channels are non-atomic) and R3 (no subdivision into sensor part, logic solver, and final element part) can easily be implemented with any of the introduced approaches. It does not matter whether the submodels represent channels or individual components. For all methods, despite the MCS approach, this results in exponentially increasing model size (state based approaches) or combinatorial expressions (combinatorial approaches), as the number of represented entities necessarily increases if channels are considered composed of substructures.

Signal replication

R2 (Component outputs should be connectable to multiple parts of the SIS) is a serious problem for the combinatorial approaches. For these, it is not generally allowed to instantiate certain components multiple times, as the stochastic dependability would be replicated incorrectly. The only valid case is expressing logical redundancies such as in, e.g., fig. 4.3. If a single sensor signal is replicated and connected to both inputs of a 1oo2

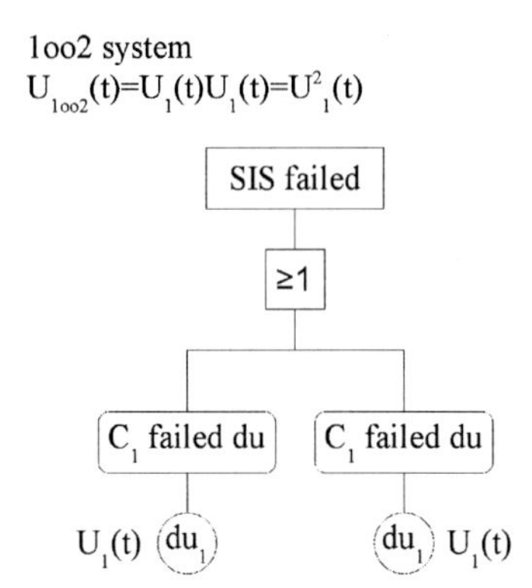

Figure 4.23: Fault tree for a replicated sensor signal

voter, a FT according to fig. 4.23 seems to be a valid result. The related unavailability would be $\mathrm{U}_1^2(t)$. But this is wrong as becomes immediately clear: in case of a component failure, both voter inputs are identically invalid, no redundancy exists. Therefore, the resulting unavailability must be $\mathrm{U}_1(t)$. This problem holds even if a general rule is introduced that requires to merge identical inputs of an OR gate as present in fig. 4.23: it is easy to construct examples that inject a failure of component C_1 at totally different points in the FT, still resulting in an erroneous result. Thus, without serious extensions to the evaluation scheme, R2 is not applicable. For state based approaches, the described problem does not exist. A state of, e.g., a MM can be interpreted as a snapshot of the overall system state at an arbitrary point in time. Each physically present component is represented by its individual element in the state vectors. The information on whether the output of any of the components is utilized multiple times in the system is not explicitly present in the MM. Instead, it is considered during the state evaluation phase, where each state is investigated with respect to whether it represents a dangerous system state or not. The problem is therefore easier to solve, as system unavailability is to be investigated considering concrete component states rather than stochastically known component states.

Maintenance groups

Another type of stochastic dependency is introduced via R7 (components should be assignable to 'maintenance groups'). Here, the restoration of a specific channel induces a repair process for all components in the related maintenance group. Combinatorial approaches as presented here are not capable of reproducing such effects. An approach to model maintenance groups with MM or MCS is basically possible. Figure 4.24 shows an

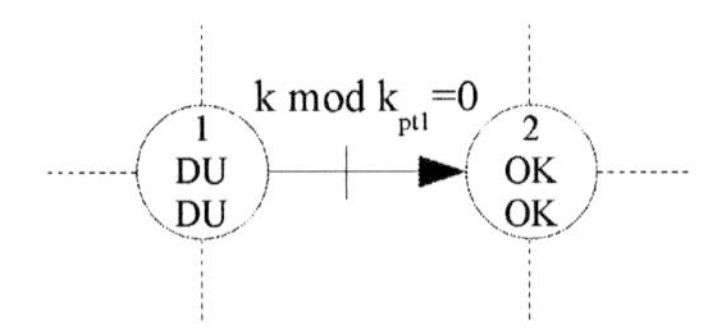

Figure 4.24: MM, implementing maintenance groups

example of how to implement these stochastical interdependencies. It shows an extract of a much larger MM. Both represented components are in a common maintenance group. In state 1, both components have failed dangerous undetected. Component 1 is proof tested regularly each k_{pt1} discrete time steps. Due to these proof tests, component 1 gets restored back to its *ok* state (transition $1 \rightarrow 2$). Since both components are in the same maintenance group, the *du* failure in component 2 gets revealed and repaired, too. Notice that this very simple model does not consider component repair times.

Diversity

R4 (individual failure rates and maintenance parameters per component), R5 (proof testability at arbitrary points in time), R6 (replaceability at arbitrary points in time), as well as R9 (partial proof testability at arbitrary points in time), can easily be implemented with all of the above mentioned approaches. The premise is that no stochastical dependencies occur requiring all sorts of proof tests and replacements deterministically scheduled. Proof tests due to the maintenance group effect can only be handled with state based approaches as explained in the previous subsection.

Multiple failure types

R8 (components may fail dangerous and safe) requires detailed consideration, as active failures need to be considered. It is not possible to simply use two isolated mathematical models and calculate safety related and operational availability separatedly from each other, since both types of failure heavily interact. A5 (Inhibition) shows that, e.g., an active failure might get inhibited by a *du* failure in a subsequentially allocated component and thus never reaches the SPLC until the passive failure is removed at the upcoming proof test. The required models therefore need to reproduce both types of failure and their potential interactions at once. At this point it becomes clear, that all approaches lack suitable evaluation methods concerning operational availability.

A 1oo2 voting algorithm requires both inputs failed *du* in order to have its output failed *du*. On the other hand, the same voter only requires a single active failure at any input in order to produce a false trip at its output, triggering, e.g., its final elements. The evaluation logic for active failures differs significantly from the one for dangerous failures as, e.g., [LR08b] derive in closed equational form (disregarding inhibition and further interdependencies). Therefore, the desired approach needs to include both basic types of failure and additionally provide two different sets of evaluation directives for calculating safety related as well as operational characteristics. In their basic form, only the state based approaches are suitable for this demand as the related calculation model mainly represents internal component states. Information on operational and safety related unavailability can be retrieved using different selection vectors, while the model itself remains the same. For combinatorial approaches, a redesign or transformation of the model is required, as the evaluation equations are integral part of the structure (in the form of AND and OR gates).

Flexible voting

A similar problem is imposed by R10 (flexible voting schemes). Here, basic combinatorial approaches provide AND , OR, and NOT gates only. Flexibility is dramatically limited. The degradation concept, i.e., changing the voting scheme in case of a pending detected sensor failure, is not applicable as it induces stochastical dependencies. The state based approaches, again, have all relevant information condensed into the selection vectors. They are therefore generally capable of providing the demanded flexibility.

Maintenance teams

None of the approaches provides out-of-the-box tools to implement different numbers of repair teams. However, MMs tend to provide more flexibility for such aspects, as, e.g., [Buk06], and lots of literature on queuing theory showed. Unfortunately, most of these approaches tend to result in additional states for the related MMs. Therefore, a true recommendation for neither of the available unavailability modeling approaches is possible.

4.6.2 Choice of method

Reviewing the previous subsection, it becomes immediately clear, that combinatorial approaches could only be used if they are extended significantly. Stochastical dependencies can be identified as the largest problem for these mathematical methods. However, their capability of graphically representing the system structure in a very catchy way is very important. The generic approach to be invented in this work, requires a formal description for outlining a SIS's structure. Chapter 5 will introduce the 'abstract safety markup language' (ASML) which is based on an FT-like graph, providing all required information.

The calculation model therefore should be state based. As markovian approaches deliver exact probabilistic results while providing good support in model optimization, and additionally seem to be most accepted and applied by safety engineers, a multiphase markov model approach is conducted from here on.

Chapter 5

Abstract safety markup language

5.1 Overview

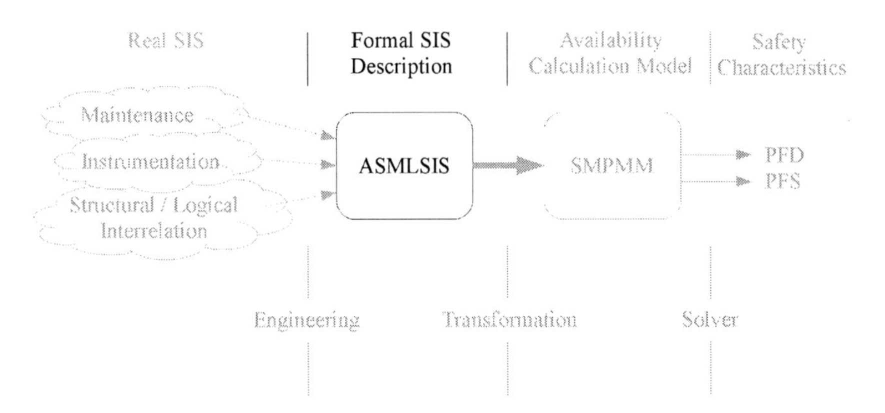

Figure 5.1: Flow diagram of the ASML approach and location of the ASMLSIS

A formal description for safety instrumented functions requires two important aspects: on the one hand the implementational details such as the devices, their wiring, voting algorithms etc. On the other hand the applied maintenance strategy, including, e.g., proof test intervals, and mission time replacements. While the first aspect typically impacts on the failure behavior of individual components and their impact on the SIS's failure behavior, the latter one deals with its restoration process. Unavailability calculations as demanded by the international safety standards and plant operators require both. The abstract safety markup language (ASML) is the entirety of mathematical structures and equations enabling for the formal description of an arbitrary safety instrumented function, then called an ASMLSIS. It is - together with the transformation provided in chapter 6 - the main contribution of this work. An ASMLSIS consists of

- a set of parameters related to the entire SIS such as a mission time and maintenance groups. This part of ASML is treated in sec. 5.4 (p. 97).
- the ASML graph, a digraph with different types of individually parameterized vertices. These vertices represent the relevant entities in the SIS: components, redundancies, and signal routing information. Its edges represent temporal and logical interdependencies among the entities. The ASML graph is introduced in sec. 5.3 (p. 77).

ASML comes with a few tools serving as a basis for the understanding of how all requirements and assumptions from sec. 2.4 (p. 24) are treated as well as a set of evaluation formulas for further treatment. These tools are

- behavioral patterns for components, describing mechanisms for the transition between component states. They reflect important assumptions of static and dynamic behavior of SISs as introduced in subsec. 2.4. The subsequent sec. 5.2 (p. 72) deals with these patterns which are neither used for calculation nor for evaluation purposes. They only facilitate the access to the ASML-related mathematical declarations.
- a set of equations for performing a logical unavailability evaluation of the ASML graph. The equations' input parameters can be retrieved from the ASML graph's vertex parameterization, as well as from ASMLSIS global parameters. The evaluation results are utilized in order to construct the selection vectors of the generic MM approach. The unavailability evaluation equations are provided in subsec. 5.3.6 (p. 89).

From fig. 5.1, the location of the ASMLSIS within the entire ASML approach can be denoted: An ASML description is created by safety engineers, typically tool based. The engineering effort is used to translate all required information about the real SIS into ASML. Section sec. 5.5 (p. 99) provides some additional guidelines on how to create valid ASML descriptions from conventional IEC 61508 structures (such as MooN votings). Afterwards, the ASMLSIS can automatically be translated into a multiphase markov model calculation model. This transformation is described in the subsequent chapter 6.

5.2 ASML global behavioral patterns

5.2.1 Component states

The possible states a component may enter during its lifetime are collected in the set of component states.

Definition 5.1 (Component states). *The set of component states $\mathcal{CS}$ is defined as*

$$\begin{aligned}\mathcal{CS} &= \mathcal{CS}_{OK} \cup \mathcal{CS}_{A} \cup \mathcal{CS}_{D} \\ &= \{OK\} \cup \{AU, AR\} \cup \{DDU, DU, DUP, DN, DDR, DR\},\end{aligned}$$

where $\mathcal{CS}_D$ is the subset of dangerous failure states, preventing the component from performing its intended safety related function. $\mathcal{CS}_A$ is the subset of active failure states, containing only failures that prevent the component from performing its intended operational function. Notice that $\mathcal{CS}_D$ and $\mathcal{CS}_A$ are mutually exclusive: $\mathcal{CS}_D \cap \mathcal{CS}_S = \emptyset$. The acronyms read as follows: active unrevealed (AU), active under repair (AR), dangerous detected unrevealed (DDU), dangerous undetected (DU), dangerous undetected partial proof test detectable (DUP), dangerous non-detectable (DN), dangerous detected under repair (DDR), dangerous under repair (DR).

A detailed description of all component states is given in subsec. 5.2.3 in the context of the component finite state machine.

5.2.2 Failure types

Definition 5.2 (Failure types). *The set of failure types $\mathcal{FT}$ is defined as*

$$\begin{aligned}\mathcal{FT} &= \mathcal{FT}_a \cup \mathcal{FT}_d \\ &= \{a\} \cup \{dd, ddx, du, dup, dn\},\end{aligned}$$

where $\mathcal{FT}_d$ is the subset of dangerous failures, preventing the component from performing its intended safety related function by leading to a component state $s \in \mathcal{CS}_D$. $\mathcal{FT}_a$ is the subset of active failures, containing only failures that prevent the component from performing its intended operational function by leading to a component state $s \in \mathcal{CS}_A$. The acronyms read as follows: active (a), dangerous detected (dd), dangerous detected with external communication (ddx), dangerous undetected (du), dangerous undetected partial proof test detectable (dup), dangerous non-detectable (dn).

As for the component states: a detailed description of all failure types is given in subsec. 5.2.3 in the context of the component finite state machine. Notice that component states are written in capital letters, while failure types are denoted in lowercase. This way the confusion of the very similar abbreviations is avoided.

5.2.3 Component finite state machine

Chapter 2 introduced a large amount of effects with relevance for the behavior of components. While the failure mechanisms are more or less easy to describe with several superimposed exponentially distributed RF processes, the according FR processes are complicated. Components may reach their functional state by either repair or replacement. The mechanisms leading to the revelation of the failure state include proof tests, partial proof tests or replacement for the considered component itself or a component in the local maintenance group etc. This work utilizes a global component behavioral pattern containing all considered effects impacting on the component's internal state. It will be given as a finite state machine which serves the only purpose of explaining the provided component behavior. It is not used explicitly in the later course of the work. Instead, the various transitions and states of the finite state machine will appear implicitly in the MPMMs generated by the transformation formulas. It is important to point out that not each transition of the finite state machine will be represented by a transition in the MPMM. Several effects are implemented as, e.g., phase transition reset procedures. As already denoted, the component finite state machine (CFSM) represents the possible behavior for *all* components. Individualization is achieved by individually parameterizing the underlying event generators responsible for state transitions. The complete parameter set explicitly describing an individual component is provided in subsec. 5.3.3 (p. 83). Notice that the CFSM reflects important design decisions.

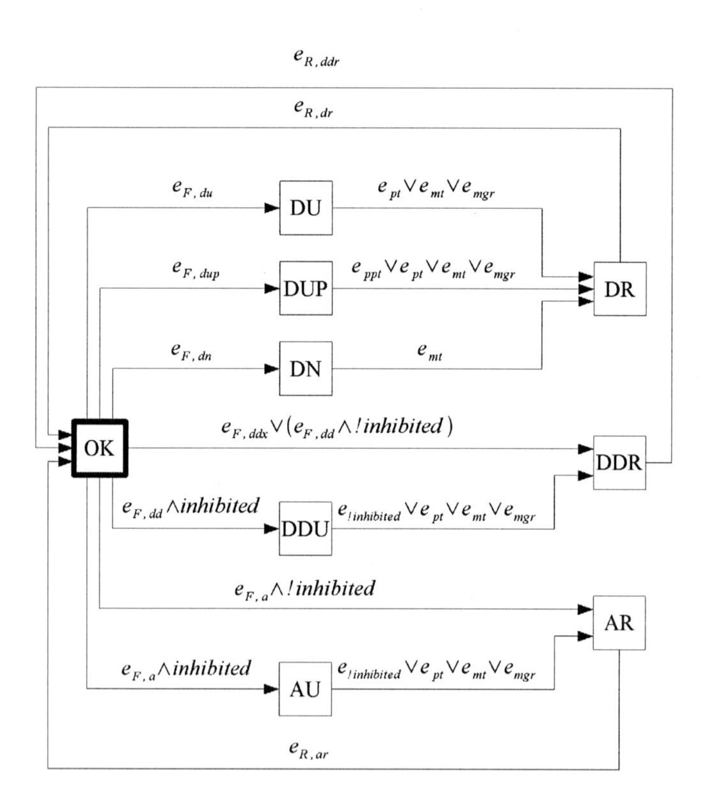

Figure 5.2: Finite state machine, representing internal component states and transitions

Subsequently, a list with explanations for all possible states, events, and utilized state variables is given. The latter ones are used to indicate whether the overall system (the SIS) is in a state with certain characteristics. These state variables change their (boolean)

value synchronized with the SIS state. The directly or indirectly reflected effects have already been described in detail in sec. 2.2 (p. 10) and sec. 2.3 (p. 19). The most important conclusions from these two sections have been condensed into the requests and assumptions from sec. 2.4 (p. 24). The belowstanding list refers to the items of said collection.

States

- **OK**: initial state of the CFSM; the component works as intended and is capable of performing its safety related as well as its operational function
- **DU**: the component suffers from a dangerous undetected (*du*) failure; it is not capable of performing its safety related function
- **DUP**: the component suffers from a dangerous undetected but partial proof test detectable (*dup*) failure; it is not capable of performing its safety related function
- **DN**: the component suffers from a dangerous non-detectable (*dn*) failure; it is not capable of performing its safety related function
- **DDU**: the component suffers from an unrevealed dangerous detected (*dd*) failure that has not yet reached an SPLC in order to have the detected failure annunciated; the component is not capable of performing its safety related function
- **AU**: the component suffers from an unrevealed active (*a*) failure that has not yet reached an SPLC or tripped a final element in order to have the active failure annunciated; the component is not capable of performing its operational function
- **DR**: the component is under repair after a dangerous (*du*, *dup*, or *dn*) failure; it is not capable of performing its safety related function
- **DDR**: the component is under repair after a dangerous detected (*dd* or *ddx*) failure; the component is not capable of performing its safety related function
- **AR**: the component is under repair after an active (*a*) failure; it is not capable of performing its operational function

Events

- $\boldsymbol{e_{F,du}}$: occurrence of a dangerous undetected (*du*) failure (A6)
- $\boldsymbol{e_{F,dup}}$: occurrence of a dangerous undetected partial proof test detectable (*dup*) failure (A8)
- $\boldsymbol{e_{F,dd}}$: occurrence of a dangerous detected (*dd*) failure (A2)
- $\boldsymbol{e_{F,ddx}}$: occurrence of a dangerous detected with external communication (*ddx*) failure (A3)

- $e_{F,dn}$: occurrence of a dangerous non-detectable (*dn*) failure (A7)
- $e_{F,a}$: occurrence of an active (*a*) failure (A4)
- $e_{R,ar}$: completion of the restoration process from the active failure repair (AR) state
- $e_{R,ddr}$: completion of the restoration process from the dangerous detected repair (DDR) state
- $e_{R,dr}$: completion of the restoration process from the dangerous repair (DR) state
- e_{pt}: initiation of a proof test for the considered component (R5)
- e_{ppt}: initiation of a partial proof test for the considered component (R9)
- e_{mt}: initiation of a replacement due to expiration of the useful lifetime for the considered component (R6)
- e_{mgr}: initiation of a component restoration process due to the restoration process completion of a different component from the same maintenance group as the considered component (R7)
- $e_{!inhibited}$: a hitherto unrevealed potentially detectable failure of the considered component (i.e., a component in state DDU or AU) becomes annunciated by, e.g., an SPLC. This causes transitions $DDU \rightarrow DDR$ or $AU \rightarrow AR$.

State variables

- ***inhibited***: a boolean variable indicating whether a potential *dd* or *a* failure of the considered component would be inhibited (A5); the variable has a distinct value for each point in time; $!inhibited$ refers to the variable evaluating as $false$

Most notable fact to be obtained from the CFSM is that neither a DD state nor an A state do exist. This results from the fact that both types of failure are either immediately detected and announced, or remain temporarily unrevealed due to inhibition effects otherwise. If immediately detected, the duration of the related FR process is clearly dominated by the repair time. Therefore, transitions related to *dd* and *a* failures which are not inhibited in the moment of occurrence, immediately lead to the respective repair states (DDR and AR), from where the OK state can be reached upon completion of the repair processes. The state sequences $OK \rightarrow DDR \rightarrow OK$ and $OK \rightarrow AR \rightarrow OK$ are thus perfect examples of RFR-IRI processes according to subsec. 3.5.1 (p. 36). If *dd* and *a* failures are inhibited in the moment of occurrence, the components change their state to DDU and AU respectively, indicating that a potentially detectable failure is contained to be announced as soon as the inhibition vanishes. This introduces a new type of RFR-DRI process, where the delay of the initiation of the underlying FR process depends on the internal state of the subsequent components. As soon as, e.g., a subsequently located *du* failure gets repaired ,i.e., $DR \rightarrow OK$, the local inhibition is resolved, the repair process initiated. The inhibition concept therefore introduces heavy interdependencies among components. The implementation of this specific feature as can be seen in the state sequences $OK \rightarrow DDU \rightarrow DDR \rightarrow OK$ and $OK \rightarrow AU \rightarrow AR \rightarrow OK$ will be given extended attention in the course of this thesis. Basically, these two cycles can be considered variations of type C RFR-DRI processes according to subsec. 3.5.4 (p. 40).

Notice that inhibition is mentioned here in the context of delaying the first annunciation of a detectable failure. It is additionally possible that a revealed *dd* failure under repair (i.e., in DDR state) gets inhibited again, as a subsequently allocated component may fail with a *du* failure. This re-inhibition does not affect the FR process of the detected failure, as a repair team is assumed to get informed in the moment of the failure's annunciation and continues with the repair process even if the diagnostic signal suddenly becomes inhibited, i.e., invisible for the repair team.

As dangerous detected failures with external communication (ddx) are considered to be not inhibitable, their occurrence always immediately initiates the related FR process. Therefore, ddx failures are treated with the state sequence $OK \rightarrow DDR \rightarrow OK$, referring to a standard RFR-IRI process according to subsec. 3.5.1 (p. 36).

Leaving event $\boldsymbol{e_{mgr}}$ out from first considerations, the state sequences $OK \rightarrow DU \rightarrow DR \rightarrow OK$, $OK \rightarrow DUP \rightarrow DR \rightarrow OK$, and $OK \rightarrow DR \rightarrow DN \rightarrow OK$ can be classified as pure RFR-DRI processes of type C according to subsec. 3.5.4 (p. 40).

Now, taking $\boldsymbol{e_{mgr}}$ into account, an additional failure revelation mechanism arises which can be considered a new variation of type C RFR-DRI process of high complexity. In contrast to the standard type C approach from subsec. 3.5.4 (p. 40), the queuing time until the initiation of the repair process depends not only on the predefined deterministic proof tests, replacements etc. , but also on having an arbitrary component in the same maintenance group undergoing a transition from a repair state back to OK. It is obvious that these two random processes compete with each other. The winning process fires $\boldsymbol{e_{mgr}}$. This behavior is similar to the superposition of multiple RF processes competing for the type of occurring failure. The difference lies in the much more complicated random variables.

5.3 ASML graph

Central object in ASML SIS descriptions is the ASML graph, a digraph with different types of attributized vertices. It visualizes the functional chain over which component failures impact on the SIS's unavailability (safety related or operational). The graph contains specific information about

- individual component failure behavior
- individual component maintenance
- structural and logical component interrelation, e.g., redundancies, and allocation

Although the ASML graph looks similar to a FT, the purpose is quite different. While FTs are utilized to directly construct unavailability functions, the ASML graph has only descriptive character. However, a set of evaluation functions is provided along with the graph for supporting the generic MPMM construction process. These functions condense specific fractions of the information contained in the ASML graph into useful characteristics. The evaluation formulas are given in subsec. 5.3.6 (p. 89).

component sequence voter

Figure 5.3: ASML graph symbols

Definition 5.3 (ASML graph). *The tuple*

$$\mathcal{ASMLG}\,(\mathcal{V},\mathcal{E})$$

is called an ASML (di)graph (ASMLG). $\mathcal{V}$ *is a set of vertices.* $\mathcal{E}$ *is a set of edges according to*

$$\mathcal{E} \subset (\mathcal{V} \times \mathcal{V})\,.$$

The interpretation of an edge $(\mathrm{e}_1, \mathrm{e}_2) \in \mathcal{E}$ *is as follows:* e_2*'s safety related/operational availability depends on* e_1*'s safety related/operational availability. Thus, the desired safety characteristics of* e_2 *depend on all vertices* $\{\mathrm{e} \mid (\mathrm{e}, \mathrm{e}_2) \in \mathcal{E}\}$.

The attributized vertices $\mathcal{V}$ *of an ASMLG for a given SIS are of three mutually different types (subsequently defined in detail):*

$$\mathcal{V} = \mathcal{COM} \cup \mathcal{SEQ} \cup \mathcal{VOT}, \textit{with } \mathcal{COM} \cap \mathcal{SEQ} \cap \mathcal{VOT} = \emptyset.$$

$\mathcal{COM}$ *is a set of components (see subsec. 5.3.3 (p. 83)),* $\mathcal{SEQ}$ *is a set of sequences (see subsec. 5.3.4 (p. 85)), and* $\mathcal{VOT}$ *is a set of voters (see subsec. 5.3.5 (p. 85)). All three vertex types are represented by specific symbols as shown in fig. 5.3.*

Some remarks on def. 5.3:

- The local parameterization for each vertex may be denoted in the graphical representation of the ASMLG or left out for simplification purposes. However, the related mathematical objects must always be fully parameterized.
- 'Voter' refers to the voting 'function', i.e., the voting algorithm, not the device that executes the algorithm.
- 'Sequence' refers to a channel structure, where multiple components are sequentially arranged. The members of a sequence forward a signal in strict temporal order.
- Both the set of vertices and set of edges, are obviously restricted, since, e.g., $\mathcal{E} \not\subseteq (\mathcal{V} \times \mathcal{V})$. A complete list of restrictions is outlined in subsec. 5.3.2 (p. 82).

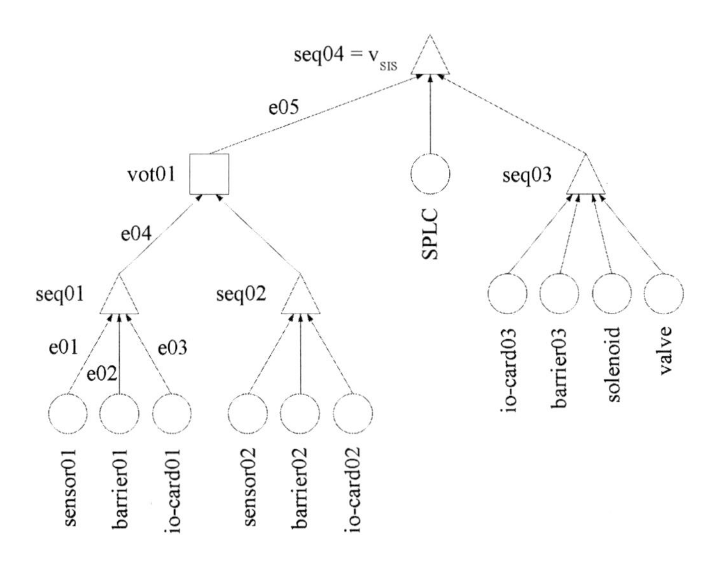

Figure 5.4: ASML graph related to the SIS in fig. 2.1 (p. 9)

Definition 5.4. *For reasons of short notation, the term 'safety related (un)availability' will be abbreviated as 's-(un)availability' from now on. 'Operational (un)availability' will be abbreviated 'o-(un)availability'.*

Figure 5.4 shows an example ASML graph. It is related to the the SIS given as wiring diagram in fig. 2.1 (p. 9). Each physically present device is outlined as an individual component (circle symbol). The sequential arrangement of both, components in sensor channels and in the final element part, is transferred into sequences (seq01, seq02, seq03, outlined as triangles). The 1oo2 voting algorithm is represented by the software voter block (vot01, square symbol). Vertices with multiple abducent edges represent signal replication (see requirement R2). Vertices with arbitrary adducent edges represent signal condensing as performed by, e.g., voting algorithms (see requirement R10). According to def. 5.3, several dependencies can be pointed out by way of example.

Edges e01, e02 and e03 indicate that the sequence seq01's availability depends on the components sensor01, barrier01, as well as io-card01 - in this order. Compare the order of components in fig. 2.1 (p. 9): sensor01 creates a signal which is afterwards forwarded by barrier01 and finally reaches io-card01. This order is reflected by the arrangement of the respective components in fig. 5.4 from left to right.

Thus, the availability of the sensor channel sensor01 → barrier01 → io-card01 depends on the availability of the contained components. The sequence's availability is representative for the availability of all three underlying components. Obviously, if all components attached to a sequence are s-available, then the sequence itself is s-available.

The next hierarchical level is denoted by the voter vot01. Its availability depends on the availability of the underlying channels seq01 and seq02 among which the voting is performed. In case of a 1oo2 voting, the vot01 redundancy subsystem is s-available, if at least one of the underlying channels (seq01 and seq02) is.

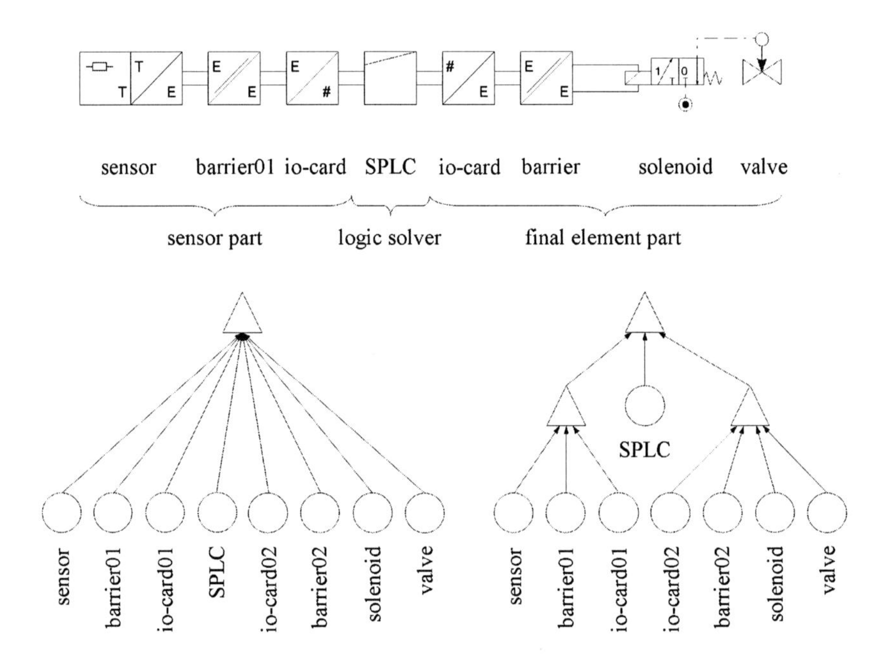

Figure 5.5: Simple SIS with purely sequential component arrangement and two possible related ASMLGs

Property 5.1 (Ambiguity of ASMLGs). *It is possible to construct different ASMLGs related to the same given SIS. As far as the interdependencies among the components are represented correctly, the transformation equations provided later on in this work will deliver identical calculation results and for the most part identical calculation models.*

Figure 5.5 shows a simple example for that. Both provided ASMLGs result in the same MM (see subsec. 8.2.6 (p. 146)). The intention for designing the ASMLG as shown in the second graph on the right is to keep a visual reference to the differentiation between sensor part, logic solver, and final element part.

5.3.1 ASMLG traversing

In order to provide flexible referencing of certain objects in the ASMLG, two important operator pairs need to be introduced. Each edge within the graph starts at a certain single vertex and ends at another single vertex. Following [Lit05], the pre and post vertices can be defined according to:

Definition 5.5 (Pre and post vertices). *If* $e = (v1, v2) \in \mathcal{E}$*, then*

$$\bullet\, e = v1 \text{ is called "pre vertex" and}$$
$$e\bullet = v2 \text{ is called "post vertex".}$$

In fig. 5.4, e.g., •e03 = io-card01, and e03• = seq01 hold. The counterparts of pre and post vertices are pre and post edges. As each vertex may have multiple pre and post edges (due to signal replication and condensation), the result of the related operator is a set of edges:

Definition 5.6 (Pre and post edges). *If* $v \in \mathcal{V}$*, then the set*

$$\bullet\, v = \{(v1, v2) \in \mathcal{E} | v2 = v\} \text{ is called "pre edges" and}$$
$$v\bullet = \{(v1, v2) \in \mathcal{E} | v1 = v\} \text{ is called "post edges".}$$

In fig. 5.4, e.g., •seq01 = {e01, e02, e03}, and seq01• = {e04} hold. With these definitions, it is possible to define edge paths. Edge paths are sequences of edges, where the post vertex of any edge in the path is the pre vertex of the subsequent edge:

Definition 5.7 (Edge path). *An arbitrary edge path p of length l from the ASMLG's set of edge paths $\mathcal{EP}$ is defined as*

$$p \in \mathcal{EP} : e_1 \to e_2 \to \cdots \to e_l,$$

where $e_i \in \mathcal{E}$ and $\forall\,(i > 1)\,(e_{i-1}\bullet = \bullet e_i)$. An edge path can be utilized using subsequent notations:

- *$p[i]$ refers to the i-th element e_i of the edge path*
- *$e \in p$ means that $\exists i\,(p[i] = e)$, i.e., edge e is a member of the edge path*
- *$|p|$ delivers the length l of the edge path p*

The set of edge paths $\mathcal{EP}$ consists of all possible edge paths constructable from the ASMLG. As the set of edges for ASMLGs is restricted (see def. 5.3), the set of edge paths is also restricted. See subsec. 5.3.2 for details.

The transformation formulas presented in chapters 6 and 7 require detailed information from the ASMLG in order to perform an optimization of the model size for increasing the calculation speed significantly. An important information is whether certain vertices of the graph are stochastically independent from each other (this expression makes sense, since vertices *are*, e.g., components having mutual stochastical relations to other components). A required criterion includes an investigation on whether the investigated vertices are connected via an edge path. It is therefore useful to define a function that searches for all edge paths interconnecting two arbitrary vertices:

Definition 5.8 (Edge path function). *The edge path function* EP *yields a set consisting of all edge paths interconnecting two arbitrary vertices* v_1 *and* v_2 *in* $\mathcal{V}$*:*

$$\mathrm{EP} : \mathcal{V} \times \mathcal{V} \mapsto \mathcal{EP} := \mathrm{EP}(v_1, v_2) = \{p \,|(\bullet p[1] = v_1) \wedge (p[|p|]\bullet = v_2)\}\,.$$

Definition 5.9 (Edge path destination function). *The edge path destination function* EPD *yields a set consisting of all edge paths leading to an arbitrary vertex* v *in* $\mathcal{V}$*:*

$$\mathrm{EPD} : \mathcal{V} \mapsto \mathcal{EP} := \mathrm{EPD}(v) = \{p \in \mathcal{EP} \,|(p[|p|]\bullet = v)\}\,.$$

5.3.2 Restrictions of ASMLGs

With these definitions it is easy to define the restrictions that have to be kept when contructing an ASMLG. Restrictions exist for the set of edges as well as for the layout of the ASMLG.

Definition 5.10 (ASMLG restrictions). *The set of edges* $\mathcal{E} \subset (\mathcal{V} \times \mathcal{V})$ *consisting of tuples* (v_1, v_2) *with* $\{v_1, v_2\} \subseteq \mathcal{V}$ *for an arbitrary ASML graph* $\mathcal{ASMLG}$ *is restricted according to*

1. $\mathcal{E} \subseteq ((\mathcal{V} \times \mathcal{V}) \setminus (\mathcal{V} \times \mathcal{COM}))$
2. $(\exists! v_{SIS} \in \mathcal{V})\,((\nexists\,(v_1, v_2) \in \mathcal{E})\,(v_1 = v_{SIS}))$

The layout of an arbitrary ASMLG is restricted according to

3. $\{p \mid \exists m \exists n\,((\bullet p[m] = p[n]\bullet) \wedge (m \leq n))\} \nsubseteq \mathcal{EP}$

Some remarks on def. 5.10:

- Restriction 1 indicates that components have outputs only. Their availability therefore only depends on internal state and not on other entities. If the edges in an ASMLG are oriented from bottom to top, then components are located at the graph's bottom. They therefore serve a purpose similar to primary events in FTs.
- Restriction 2 outlines that a single entity in the whole ASMLG is allowed / required to not have an abducent edge, rendering this entity the SIS's unavailability output v_{SIS} (see sequence seq4 in fig. 5.4). Its unavailability dictates the overall system unavailability. The restriction ensures that v_{SIS} is implicitly defined in each ASMLG.
- Restriction 3 refers to edge loops which are forbidden. Otherwise, algebraic loops would occur in the evaluation formulas for ASMLGs. A more detailed look to this problem will be given in subsec. 5.3.6 (p. 89).
- As restriction 3 guarantees loop freeness of ASMLGs, arbitrary edge paths as, e.g., obtained via the edge path function (def. 5.8), are also loop free.

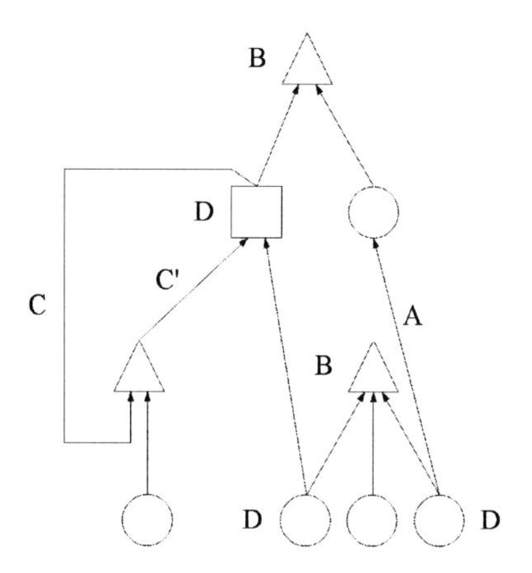

Figure 5.6: ASML graph violating all three restrictions from def. 5.10

Figure 5.6 shows a violation of all three restrictions introduced in def. 5.10 for illustrative purposes. The edge (v_1, v_2) marked with 'A' has $v_2 \in \mathcal{COM}$, violating restriction 1. Two sequences are denoted as 'B'. Both do not have any abducent edge and therefore violate restriction 2. The edge at 'C' disregards restriction 3, as an edge path C $\rightarrow$ C' exists with $\bullet$C = C'$\bullet$. Opposing to that, all signal replications (multiple abducent edges at vertices denoted as 'D') are tolerable.

5.3.3 Components

Component vertices in ASMLGs are parameterized with information about individual maintenance and failure behavior.

Definition 5.11 (Component). *An arbitrary component c_i in the set of components $\mathcal{COM}$ is a six-tuple*

$$c_i = \mathcal{COMPONENT}\left(\boldsymbol{\lambda}_i, \boldsymbol{\mu}_i, \sigma_i, \mathcal{PT}_i, \mathcal{PPT}_i, \mathcal{MT}_i\right).$$

$\boldsymbol{\lambda}_i$ is a vector of failure rates for R different failure types according to

$$\boldsymbol{\lambda}_i = \left(\lambda_{dd,i}\ \lambda_{ddx,i}\ \lambda_{du,i}\ \lambda_{dup,i}\ \lambda_{dn,i}\ \lambda_{a,i}\right)^\top.$$

$\boldsymbol{\mu}_i$ is a repair rate vector according to

$$\boldsymbol{\mu}_i = \left(\mu_{ddr,i}\ \mu_{dr,i}\ \mu_{ar,i}\right)^\top.$$

σ_i is a component state function (see def. 5.12). $pt_j \in \mathcal{PT}_i \subset \mathbb{R}_+$ denote all points in time a proof test is initiated for the component c_i. $ppt_j \in \mathcal{PPT}_i \subset \mathbb{R}_+$ denote all points in time a partial proof test is conducted. $mt_j \in \mathcal{MT}_i \subset \mathbb{R}_+$ denote all points in time a component replacement due to extended mission time is conducted.

Most of the parameters, such as the proof test time steps or failure and repair rates are easy to understand and closely related to the events denoted for the CFSM in subsec. 5.2.3 (p. 73) (see also def. 5.2 (p. 73) for the failure types). However, the idea of the component state function $\sigma(t)$ needs to be explained.

Definition 5.12 (Component state function). *The component state function $\sigma(t)$ yields the current state of a component for each point in time. The possible component states are defined in the component state set $\mathcal{CS}$ (see def. 5.1).*

$$\sigma : \mathbb{R}_+ \mapsto \mathcal{CS} := \mathbb{R}_+ \mapsto \{AU, AR, OK, DDU, DU, DUP, DN, DDR, DR\}\,.$$

It is important to point out, that an explicit mathematical formulation for $\sigma(t)$ is generally not possible, as it is subject to heavy interdependencies with other components and is additionally influenced by the SIS's structure as defined in the ASMLG. However, it is sufficient to postulate the *existence* of such a function. Section sec. 5.3.6 (p. 89) as well as sec. 6.3 (p. 107) will give a more detailed insight into the application of the component state function.

Example 5.1 (Generic solenoid parameterization). A generic solenoid may serve as an example for the parameterization of a safety component. Basic failure rates are obtained from [exi].

Table 5.1: Exemplary parameterization for a generic solenoid (rates in h^{-1}, times in y)

Parameter	Value(s)
$\boldsymbol{\lambda}$	$(0 \quad 0 \quad 6\times10^{-9} \quad 574\times10^{-9} \quad 0 \quad 1000\times10^{-9})^{\mathsf{T}}$
$\boldsymbol{\mu}$	$(1.25\times10^{-1} \quad 1.25\times10^{-1} \quad 1.25\times10^{-1})^{\mathsf{T}}$
$\mathcal{PT}$	$\{1.0, 2.0\}$
$\mathcal{PPT}$	$\{0.5, 1.5, 2.5\}$
$\mathcal{MT}$	$\{\}$

The data from table 5.1 are ASML data for a solenoid with behavior as follows: $\lambda_a = 1000\,\mathrm{FIT}$ refers to active failures, causing the solenoid to actuate the subsequently allocated final element unintendedly. The failure rate $\lambda_{dup} = 574\,\mathrm{FIT}$ refers to dangerous undetected failures that can be revealed by partial stroke tests as well as regular proof tests (i.e., full-stroke-tests). The missing 6 FIT are assigned to λ_{du}, i.e., the failure rate related to dangerous undetected failures, which can only be revealed during regular proof tests. Proof tests are conducted at the end of the first and second year after commissioning. Partial proof tests are conducted annually with an offset of six months with respect to the proof tests. The solenoid is never replaced.

The example shows that the observation time for the SIS is of major interest since it dictates, e.g., the amount of conducted proof test that have to be specified in a component's set $\mathcal{PT}$. As the observation time, i.e., the SIS's mission time, is a parameter related to the whole safety loop rather than to a single component, it is introduced in sec. 5.4 (p. 97), together with the rest of SIS-global parameters.

5.3.4 Sequences

Sequences define the order of N vertices that are consecutively involved in the process of forwarding a trip signal. They are characterized by one single parameter:

Definition 5.13 (Sequence). *An arbitrary sequence s_i in the set of sequences $\mathcal{SEQ}$ is a one-tuple*

$$s_i = \mathcal{SEQUENCE}\left(\boldsymbol{iv}_i\right),$$

specifying the temporal order of vertices contributing to the forwarding process of a trip signal with their respective availabilities. $\boldsymbol{iv}_i$ with $\boldsymbol{iv}_i[j] \in \mathcal{V}$ is a vector of input vertices, specifying the order of all N involved vertices. The length of this vector corresponds to the number of pre edges for the considered sequence:

$$|\boldsymbol{iv}_i| = |\bullet s_i| = N.$$

The pre vertices of the edges $\bullet s_i$ are the elements contained in the input vertex vector:

$$\{v1 \mid (v1, v2) \in \bullet s_i\} = \{\boldsymbol{iv}_i[1], \boldsymbol{iv}_i[2], \ldots, \boldsymbol{iv}_i[N]\}.$$

$\boldsymbol{iv}[1]$ is the temporarily first, and $\boldsymbol{iv}[N]$ the temporarily last required vertex. The order of input vertices corresponds to the order of the considered vertices in ASMLGs from left to right (if not explicitly defined otherwise).

Typically, sequences in ASML describe structures that resemble classical channel structures as generally assumed in IEC 61511 and IEC 61508.

Example 5.2 (Parameterization for sequences in fig. 5.4). In fig. 5.4 (p. 79), four sequences seq01 to seq04 are denoted. The related input vertex vectors can be deduced from the underlying wiring diagram in fig. 2.1 (p. 9) as well as from the arrangement of the vertices and edges in the related ASMLG from left. Thus,

$$\begin{aligned}
\boldsymbol{iv}_{\text{seq01}} &= (\text{sensor01 barrier01 io-card01})^\top \\
\boldsymbol{iv}_{\text{seq02}} &= (\text{sensor02 barrier02 io-card02})^\top \\
\boldsymbol{iv}_{\text{seq03}} &= (\text{io-card03 barrier03 solenoid valve})^\top \\
\boldsymbol{iv}_{\text{seq04}} &= (\text{vot01 SPLC seq03})^\top
\end{aligned}$$

can be derived. A potential trip signal is therefore generated, e.g., at sensor01, reaches barrier01 afterwards, and finally gets to io-card01. In that order.

5.3.5 Voters

Definition 5.14 (Voter). *An arbitrary voter v_i in the set of voters $\mathcal{VOT}$ is a two-tuple*

$$v_i = \mathcal{VOTER}\left(\boldsymbol{vs}_i, \boldsymbol{iv}_i\right),$$

specifying the redundancy among the vertices contributing to the forwarding process of a trip signal with their respective availabilities. $\boldsymbol{iv}_i$ *with* $\boldsymbol{iv}_i[j] \in \mathcal{V}$ *is a vector of input vertices, specifying an explicit individual index for all N involved vertices. The length of this vector corresponds to the number of pre edges for the considered voter:*

$$|\boldsymbol{iv}_i| = |\bullet v_i| = N.$$

The pre vertices of the edges $\bullet v_i$ *are the elements contained in the input vertex vector:*

$$\{v1 \mid (v1, v2) \in \bullet v_i\} = \{\boldsymbol{iv}_i[1], \boldsymbol{iv}_i[2], \ldots, \boldsymbol{iv}_i[N]\}.$$

$\boldsymbol{vs}_i$ *is a vector of voting schemes, i.e., voting algorithms (1oo2 is an exemplary voting algorithm; see def. 5.15), of length* $N + 1$. $\boldsymbol{vs}_i[j]$ *is the active voting scheme as long as at* $j - 1$ *voter inputs a dangerous failure is pending. The index of the input vertices corresponds to the order of the considered vertices in ASMLGs from left to right (if not explicitly defined otherwise).*

Voters operate on a set of inputs corresponding to the vertices connected to the considered voter via edges in ASMLGs. The vector of input vertices $\boldsymbol{iv}$ specifies an explicit index for each input vertex which will be reused later in the specification for the voting schemes. $\boldsymbol{vs}$ contains one voting scheme for each possible degradation step. As outlined in subsec. 2.2.2 (p. 14) and demanded in A13, a degradation may be performed whenever another *dd* failure is detected at one of the voter inputs. A voter as defined in def. 5.14 cannot differentiate explicit failed channels, i.e., works on the total number of failures only. A voter with two input channels and a failure in the first channel is in the same internal state as with a failure in the second one. The maximum number of degradations therefore corresponds to the total number of inputs. $N + 1$ voting schemes need to be provided as one scheme is required while no detected failure is pending. However, not all N degradation steps must result in a change of the current voting algorithm. The examples following the subsequent definitions will show how to specify a voting scheme such that the related voter behaves statically.

Notice that a voter obviously needs information about the amount of *dd* failures at its inputs. This information goes beyond s- and o-availability. In subsec. 5.3.6 (p. 89), the evaluation formulas will denote the *dd* failure carrier functions that serve the purpose of delivering this information.

Definition 5.15 (Voting scheme). *A voting scheme* $\mathcal{VS}$ *is a set of shutdown combination vectors, representing a static voting algorithm, according to*

$$\mathcal{VS} = \{\boldsymbol{sdc}_1, \boldsymbol{sdc}_2, \ldots\}.$$

Similar to the concept of minimal cut-sets in RBDs or FTs, each shutdown combination vector describes a combination of voter inputs that have to be s-available in order to have the considered voter vertex itself s-available:

Definition 5.16 (Shutdown combination vector). *A shutdown combination (SDC) vector is defined as*

$$\boldsymbol{sdc} = (\boldsymbol{sdc}[1] \ \dots \ \boldsymbol{sdc}[N])^\top \ \ \textit{with } \boldsymbol{sdc}[i] \in \{true, *\}\,.$$

$\boldsymbol{sdc}[i] = true$ *denotes, that the i-th voter input vertex (i.e.,* $\boldsymbol{iv}[i]$*) must be s-available in case of a demand.* $\boldsymbol{sdc}[i] = *$ *denotes that the i-th input might be s-available or not (i.e., representing a 'don't care' assignment).*

Notice that the redundancy specification is made from the point of view of s-availability. A specification of voting schemes for operational availability is *not* required. These are automatically derived from the specifications according to defs. def. 5.15 and def. 5.16. This means, a voting scheme representing a 1oo2 voting algorithm as specified by IEC 61508, needs not to be declared a 2oo2 voting with respect to o-availability (a single spurious trip of the inputs leads to operational unavailability).

The shutdown combination concept has been introduced in [GLS08b] and is extended here to satisfy the requirements of a system with dynamic voting.

Two short examples shall demonstrate the application of voting schemes:

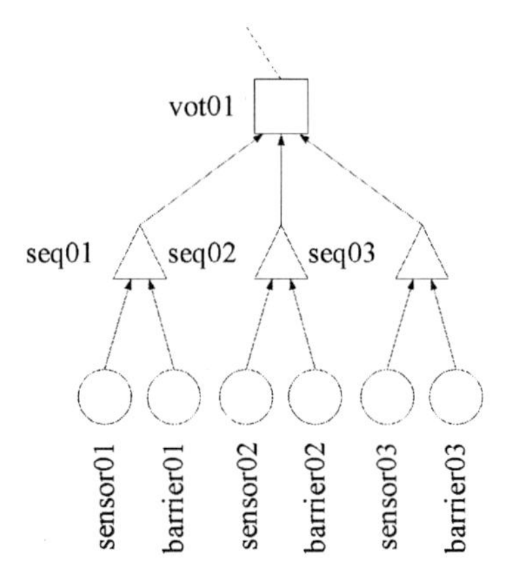

Figure 5.7: ASMLG for a 2oo3 sensor part with degradation

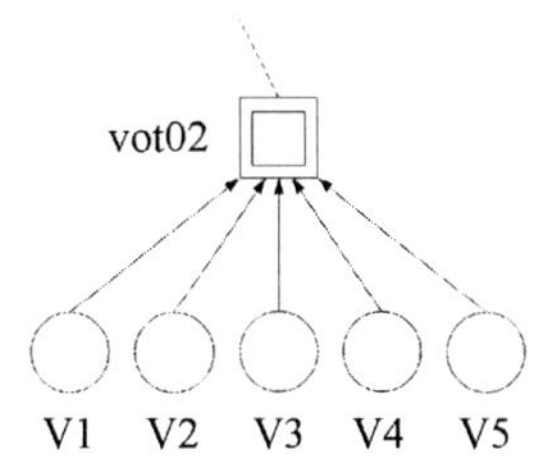

Figure 5.8: ASML hardware voter with non-standard redundancy

Example 5.3 (Parameterization for the voter in fig. 5.7). The ASMLG in fig. 5.7 refers to a SIS's sensor part with three sensor channels, each consisting of a sensor device as well as a barrier. The voting algorithm performs a 2oo3 voting scheme. Two degradation steps are implemented: If a dangerous failure is detected at any of the voter inputs, the related channel is ignored and the algorithm switches to a 1oo2 scheme. If another failure is detected while the first degradation is active, the SIS's final element part is immediately triggered. In the (unlikely) case of three detected failures, the forced trip remains active, i.e., second and third degradation scheme are identical.

The voter's inputs are specified by the vector of input vertices according to

$$\boldsymbol{iv} = (\text{seq01 seq02 seq03})^\top .$$

Since the number of voter inputs is $N = 3$, four voting schemes need to be specified. The first one refers to the 2oo3 voting which is active as long as no detected failure is pending:

$$\boldsymbol{vs}[1] = \left\{ (true\, true\, *)^\top , (true\, *\, true)^\top , (*\, true\, true)^\top \right\}.$$

Each possible combination of two s-available sensor channels enables the voter to perform as intended. Notice that the case $(true\, true\, true)^\top$ is included indirectly via the asterisk symbols. After the first degradation, the voter runs a 1oo2 scheme, ignoring the erroneous sensor channel:

$$\boldsymbol{vs}[2] = \left\{ (true\, *\, *)^\top , (*\, true\, *)^\top , (*\, *\, true)^\top \right\}.$$

Remember that a specific failed channel cannot always explicitly be denoted in voting schemes: let us assume that a *dd* failure occurs in the first sequence seq01 at an arbitrary point in time. Then $(true\, *\, *)^\top$ can definitely not be satisfied. The other two shutdown combination vectors can be written as $(false\, true\, *)^\top , (false\, *\, true)^\top$. This represents a voting scheme where either input vertex 2 or 3 needs to be s-available. This exactly corresponds to a 1oo2 voting.

After the occurrence of a second detected failure, the voter triggers the final element part (not shown in the related figure):

$$\boldsymbol{vs}[3] = \boldsymbol{vs}[4] = \left\{ (*\, *\, *)^\top \right\}.$$

The provided shutdown combination vector is to be interpreted as follows: no matter what the actual states of the voter inputs are, the voter is s-available as it is in its safe state, i.e., has forwarded a trip signal.

Example 5.4 (Parameterization for the voter in fig. 2.16)**.** In fig. 2.16 (p. 21), an exemplary final element part with an uncommon hardware redundancy scheme is depicted. The related ASMLG is shown in fig. 5.8. The description of combinations leading to a safe system state as provided there can immediately be transformed into a suitable voting scheme. The vector of input vertices is

$$\boldsymbol{iv} = (\text{V1 V2 V3 V4 V5})^\top .$$

As the safety task is to prevent the medium flow from left to right, several combinations of s-available final element channels are possible. These can immediately be transformed into shutdown combination vectors [GLS08b]. As structural voting does not allow for degrading, all voting schemes are identical:

$$\begin{aligned} \boldsymbol{vs}[1] &= \boldsymbol{vs}[2] = \ldots = \boldsymbol{vs}[6] \\ &= \left\{ (*\, *\, *\, *\, true)^\top , (*\, *\, true\, true\, *)^\top , (true\, true\, *\, true\, *)^\top \right\}. \end{aligned}$$

5.3.6 Evaluation of ASMLGs

The evaluation equations provided in this subsection will allow for making a decision on whether a SIS described by a fully parameterized ASMLG is s- or o-unavailable for a given configuration of all component state functions. In the course of this work the evaluation functions will be of great importance for determining selection vectors for the automatically constructed MM (see, e.g., subsec. 6.3.2 (p. 109)). The provided set of formulas works on generic ASML graphs of arbitrary structure (assuming that restrictions from subsec. 5.3.2 (p. 82) are met). The graph has to be evaluated recursively by investigating the SIS's output vertex v_{SIS} first. The unavailability of this vertex is representative for the whole SIS. Remember that the unavailability of an arbitrary vertex $v \in \mathcal{V}$ depends on its type, its local parameterization as well as on the unavailability of all vertices connected to the considered vertex via edges.

Since unavailability evaluation is vertex type dependent, a set of formulas has to be provided for components, sequences, as well as for voters. In order to simplify the general recursive traversing through the ASMLG, a generalized set of equations is to be introduced first. It serves as a kind of interface to the vertex type individual equations which are presented afterwards.

Definition 5.17 (Generalized ASMLG evaluation). *The generalized set of evaluation functions for arbitrary ASMLG vertices $v \in \mathcal{V}$ is*

$$\mathrm{DDC} : (\mathcal{V} \times \mathbb{R}_+) \mapsto \mathbb{N}_0$$

$$:= \mathrm{DDC}(v,t) = \begin{cases} \mathrm{DDC_{COM}}(v,t) & \textit{if } v \in \mathcal{COM}, \\ \mathrm{DDC_{SEQ}}(v,t) & \textit{if } v \in \mathcal{SEQ}, \\ \mathrm{DDC_{VOT}}(v,t) & \textit{if } v \in \mathcal{VOT}. \end{cases}$$

$$\mathrm{PFDC} : (\mathcal{V} \times \mathbb{R}_+) \mapsto \{true, false\}$$

$$:= \mathrm{PFDC}(v,t) = \begin{cases} \mathrm{PFDC_{COM}}(v,t) & \textit{if } v \in \mathcal{COM}, \\ \mathrm{PFDC_{SEQ}}(v,t) & \textit{if } v \in \mathcal{SEQ}, \\ \mathrm{PFDC_{VOT}}(v,t) & \textit{if } v \in \mathcal{VOT}. \end{cases}$$

$$\mathrm{PFSC} : (\mathcal{V} \times \mathbb{R}_+) \mapsto \{true, false\}$$

$$:= \mathrm{PFSC}(v,t) = \begin{cases} \mathrm{PFSC_{COM}}(v,t) & \textit{if } v \in \mathcal{COM}, \\ \mathrm{PFSC_{SEQ}}(v,t) & \textit{if } v \in \mathcal{SEQ}, \\ \mathrm{PFSC_{VOT}}(v,t) & \textit{if } v \in \mathcal{VOT}. \end{cases}$$

The PFD contribution function $\mathrm{PFDC}(v,t)$ *denotes whether vertex v is s-unavailable and thus contributes to the SIS's safety related unavailability at time t. The PFS contribution function* $\mathrm{PFSC}(v,t)$ *is true, if the considered vertex is o-unavailable and thus contributes to a spurious trip state, where the SIS's predefined safety function has successfully but unintendedly been triggered. The dd failure carrier function* $\mathrm{DDC}(v,t)$ *delivers a* 1 *if the considered vertex provides information about an underlying self revealing (dd or a) failure and* 0 *if it does not.*

All functions related to component evaluation (i.e., $\mathrm{DDC_{COM}}$, $\mathrm{PFDC_{COM}}$, $\mathrm{PFSC_{COM}}$*) are defined in def. 5.18. All functions related to sequence evaluation (i.e.,* $\mathrm{DDC_{SEQ}}$, $\mathrm{PFDC_{SEQ}}$, $\mathrm{PFSC_{SEQ}}$*) are defined in def. 5.19. All functions related to voter evaluation (i.e.,* $\mathrm{DDC_{VOT}}$, $\mathrm{PFDC_{VOT}}$, $\mathrm{PFSC_{VOT}}$*) are defined in def. 5.20.*

The set of evaluation equations for components c_i depends only on the internal state of the components as provided by the related component state functions σ_i, as components do not have input vertices their local unavailability may depend on.

Definition 5.18 (Component evaluation). *The set of evaluation functions for arbitrary components* $c_i \in \mathcal{COM}$ *is*

$$\mathrm{DDC_{COM}}(c_i, t) = \begin{cases} 1 & \text{if } \sigma_i(t) \in \{DDU, DDR\}, \\ 0 & \text{else.} \end{cases}$$

$$\mathrm{PFDC_{COM}}(c_i, t) = \begin{cases} true & \text{if } \sigma_i(t) \in \mathcal{CS}_D, \\ false & \text{else.} \end{cases}$$

$$\mathrm{PFSC_{COM}}(c_i, t) = \begin{cases} true & \text{if } \sigma_i(t) \in \mathcal{CS}_A, \\ false & \text{else.} \end{cases}$$

$\mathrm{DDC_{COM}}$ *maps* $(\mathcal{COM} \times \mathbb{R}_+) \mapsto \mathbb{N}_0$*. All others map* $(\mathcal{COM} \times \mathbb{R}_+) \mapsto \{true, false\}$*. An arbitrary component* c_i *is s-available at time* t_0*, if* $!\,\mathrm{PFDC_{COM}}(c_i, t_0)$*. o-availability follows accordingly.*

Some remarks on def. 5.18:

- $\mathrm{PFDC_{COM}}$ returns a contribution to the SIS's PFD if the component is in a state where it is not capable of performing its intended safety related function. These states (DDU, DU, DUP, DN, DDR, DR) are classified as dangerous states, i.e., elements of $\mathcal{CS}_D$ according to def. 5.1 (p. 72).
- Accordingly, $\mathrm{PFSC_{COM}}$ returns a PFS contribution, as long as the component is in *AU* or *AR* state (i.e., elements of $\mathcal{CS}_A$). Both of these states encode operational unavailability.

Definition 5.19 (Sequence evaluation). *The set of evaluation functions for arbitrary sequences* $s_i \in \mathcal{SEQ}$ *is*

$$\begin{aligned}&\mathrm{DDC_{SEQ}}(s_i, t)\\ &= \begin{cases} 1 & \text{if } \exists m\, ((\mathrm{DDC}(\boldsymbol{iv}_i[m], t) \neq 0) \\ & \quad \wedge \forall\, (n > m)\, (!\,\mathrm{PFDC}(\boldsymbol{iv}_i[n], t) \wedge !\,\mathrm{PFSC}(\boldsymbol{iv}_i[n], t)))\,, \\ 0 & \text{else.} \end{cases}\end{aligned}$$

$$\begin{aligned}&\mathrm{PFDC_{SEQ}}(s_i, t)\\ &= \begin{cases} true & \text{if } \exists m\, (\mathrm{PFDC}(\boldsymbol{iv}_i[m], t) \wedge \forall\, (n > m)\, (!\,\mathrm{PFSC}(\boldsymbol{iv}_i[n], t)))\,, \\ false & \text{else.} \end{cases}\end{aligned}$$

$$\text{PFSC}_{\text{SEQ}}(s_i, t) = \begin{cases} true & \textit{if } \exists m \, (\text{PFSC}(\boldsymbol{iv}_i[m], t) \wedge \forall \, (n > m) \, (! \, \text{PFDC}(\boldsymbol{iv}_i[n], t))) \, , \\ false & \textit{else.} \end{cases}$$

DDC_{SEQ} *maps* $(\mathcal{SEQ} \times \mathbb{R}_+) \mapsto \mathbb{N}_0$. *All others map* $(\mathcal{SEQ} \times \mathbb{R}_+) \mapsto \{true, false\}$. *An arbitrary sequence* s_i *is called s-available at time* t_0, *if* $! \, \text{PFDC}_{\text{SEQ}}(s_i, t_0)$. *o-availability follows accordingly.*

Some remarks on def. 5.19:

- A dangerous detected failure at vertex input with index m can only be forwarded if for all input indices $n > m$ the related vertices are capable of forwarding this information correctly. This is the case if for all $n > m$ the vertices evaluate as $\text{PFDC}(\boldsymbol{iv}_i[n], t) = false$ and $\text{PFSC}(\boldsymbol{iv}_i[n], t) = false$. Hence, the denoted equation for DDC_{SEQ}.
- A sequence contributes to the SIS's PFD via PFDC_{SEQ}, if any input of the sequence is s-unavailable, i.e., if exists an input vertex with index m and $\text{PFDC}(\boldsymbol{iv}_i[m], t) = true$ and all subsequent input vertices with index $n > m$ of the sequence are o-available, i.e., $\text{PFSC}(\boldsymbol{iv}_i[n], t) = false$.

Definition 5.20 (Voter evaluation). *The current voting scheme* $\mathcal{VS}_{cur}$ *at time* t *is*

$$\mathcal{VS}_{cur} = \boldsymbol{vs}_i[ddcnt + 1], \ \textit{with } ddcnt = \sum_{j=1}^{|\boldsymbol{iv}_i|} \text{DDC}(\boldsymbol{iv}_i[j], t).$$

The set of evaluation functions for arbitrary voters $v_i \in \mathcal{VOT}$ *is defined as*

$$\text{DDC}_{\text{VOT}}(v_i, t) \equiv 0.$$

$$\text{PFDC}_{\text{VOT}}(v_i, t) = \begin{cases} false & \textit{if } \exists \, (\boldsymbol{sdc} \in \mathcal{VS}_{cur}) \, (\forall j \, (!\boldsymbol{sdc}[j] \vee (\boldsymbol{sdc}[j] \wedge ! \, \text{PFDC}(\boldsymbol{iv}_i[j], t)))) \, , \\ true & \textit{else.} \end{cases}$$

$$\text{PFSC}_{\text{VOT}}(v_i, t) = \begin{cases} true & \textit{if } \exists \, (\boldsymbol{sdc} \in \mathcal{VS}_{cur}) \, (\forall j \, (!\boldsymbol{sdc}[j] \vee (\boldsymbol{sdc}[j] \wedge \text{PFSC}(\boldsymbol{iv}_i[j], t)))) \, , \\ false & \textit{else.} \end{cases}$$

DDC_{VOT} *maps* $(\mathcal{VOT} \times \mathbb{R}_+) \mapsto \mathbb{N}_0$. *All others map* $(\mathcal{VOT} \times \mathbb{R}_+) \mapsto \{true, false\}$. *An arbitrary voter* v_i *is called s-available at time* t_0, *if* $! \, \text{PFDC}_{\text{VOT}}(v_i, t_0)$. *o-availability follows accordingly.*

Some remarks on def. 5.20:

- $\mathcal{VS}_{cur} = \boldsymbol{vs}_i[ddcnt + 1]$ has the additional $+1$, as vector elements are denoted 1-based, i.e., the first vector element of $\boldsymbol{a}$ is $\boldsymbol{a}[1]$. As $ddcnt = 0$ is possible, the $+1$ is required.

- A voter is not capable of forwarding a dangerous detected failure information, therefore $\mathrm{DDC}_{\mathrm{VOT}}(v_i, t) \equiv 0$. As dangerous detected failures are assumed to be announced at voters, forwarding does not make sense.

- The current voting scheme at any point in time depends on the number of pending detected failures at the voter's inputs. This number *ddccnt* can be retrieved by iterating over all voter inputs and evaluating the related *dd* failure carrier functions: $ddccnt = \sum_{j=1}^{|\boldsymbol{iv}_i|} \mathrm{DDC}(\boldsymbol{iv}_i[j], t)$, where $\mathrm{DDC}(\boldsymbol{iv}_i[j], t) \in \{0, 1\}$. For simple SIS sensor system parts, *ddccnt* equals the number of sensor channels containing a non-inhibited dangerous detected failure.

- A voter contributes to the SIS's PFD if it is not capable of finding a valid shutdown combination vector $\boldsymbol{vs}$ among its current voting scheme $\mathcal{VS}_{cur}$ which has for each $\boldsymbol{vs}[j] = 1$ a corresponding input vertex which is s-available, i.e., $\mathrm{PFDC}(\boldsymbol{iv}_i[j], t) = false$.

- A voter contributes to the SIS's PFS if there exists a valid shutdown combination vector $\boldsymbol{vs}$ among its current voting scheme $\mathcal{VS}_{cur}$ which has for each $\boldsymbol{vs}[j] = 1$ a corresponding input vertex which is o-unavailable with respect to the PFS, i.e., $\mathrm{PFSC}(\boldsymbol{iv}_i[j], t) = true$.

From definitions def. 5.17 to def. 5.20 an important property of the evaluation formulas can be derived:

Property 5.2 (Mutual exclusion of PFD and PFS contribution). *For each vertex $v \in \mathcal{V}$ in an ASMLG at an arbitrary point t_0 in time*

$$\mathrm{PFDC}(v, t_0) = true \rightarrow \mathrm{PFSC}(v, t_0) = false$$
$$\mathrm{PFSC}(v, t_0) = true \rightarrow \mathrm{PFDC}(v, t_0) = false$$

holds. It is important to point out, that it is anyways possible to have $\mathrm{PFDC}(v, t_0) = \mathrm{PFSC}(v, t_0) = false$.

This property reflects the natural perception that arbitrary components or subsystems of a SIS cannot be in safe and dangerous state at the same time.

5.3.7 Static evaluation of ASMLGs

Up to now, all evaluation formulas are functions of time. The reason for this lies in the component state functions $\sigma_i(t)$. If these were known to the engineer, the (un)availability of the SIS could be calculated for arbitrary points in time. Unfortunately, this would require explicit a priori knowledge about the outcome of all underlying relevant random variables and processes. Since this is impossible, a different approach is chosen to model the contained information. The previously discussed markovian approaches implicitly encode $\sigma_i(t)$ in their transition matrix and phase transitions. It is thus possible to retrieve a probability for each possible outcome of $\sigma_i(t)$ at any point in time. In other words:

instead of knowing the explicit state of a component at t_0, it is possible to obtain the probability for the component being in state OK, DU, DDU etc. at time t_0.

However, the evaluation formulas as introduced in the previous subsection serve a very important purpose: each state space vector of a markov model represents a specific configuration of component states, e.g., $\boldsymbol{s} = (OK\ DU\ DDR)^\top$. This state could be entered by the SIS at an unknown point in time $\tilde{t}$. For the selection vector construction of the MM it is important to find out whether the SIS is s-functional or not when in said state. This can be achieved by performing an ASML evaluation according to subsec. 5.3.6 for the fictive point in time $\tilde{t}$, where all component state functions have values according to state space vector $\boldsymbol{s}$.

Definition 5.21 (Static ASML Graph Evaluation). *A static evaluation of an ASMLG is performed by postulating the existence of a point $\tilde{t}$ in time, for which all component state functions $\sigma_i(\tilde{t})$ have well-known values. These values are specified in a component state configuration vector $\tilde{\boldsymbol{\sigma}}$ of size $|\mathcal{COM}| \times 1$ with $\tilde{\boldsymbol{\sigma}}[i] \in \mathcal{CS}$ such that*

$$\tilde{\boldsymbol{\sigma}} \mathrel{\widehat{=}} \left(\sigma_1(\tilde{t})\ \sigma_2(\tilde{t})\ \ldots\ \sigma_{|\mathcal{COM}|}(\tilde{t})\right)^\top .$$

For a given component state configuration vector, an ASMLG evaluation can be performed, since all time dependent function outputs $\sigma_i(\tilde{t})$ are specified by the given vector. A markov state space vector $\boldsymbol{s} \in \mathcal{S}_{MM}$ may serve as a component state configuration vector.

The subsequent examples illustrate the interaction of the different sets of equations for the evaluation of a complete ASMLG. The respective visualizations are visually extended in order to provide more insight into the underlying evaluation steps. A vector gets attached to each vertex $v \in \mathcal{V}$ indicating the result of DDC$(v, \tilde{t})$, PFDC$(v, \tilde{t})$, and PFSC$(v, \tilde{t})$. For simplifying the visualization, $true$ is depicted as 1 and $false$ as 0. Additionally, the configuration of $\sigma_i(t)$ for the considered point in time $t = \tilde{t}$ is depicted inside the related circle symbol for each component c_i.

Example 5.5 (Component state configuration 1 for SIS from fig. 2.1 (p. 9)). The first two examples continue the investigation for the SIS shown in fig. 2.1 (p. 9). It consists of two PT100 temperature sensor channels. Each channel consists of the sensor element, a barrier, as well as an io-card connecting the channel to an SPLC. The SPLC runs a dynamic voting on the sensor channels. It starts with a 1oo2 voting and performs an emergency shutdown of the final elements if any failure is detected among the channels. This behavior can be described as an ASML voter according to

$$\begin{aligned}
&\text{vot01} = (\boldsymbol{iv}_{\text{vot01}}, \boldsymbol{vs}_{\text{vot01}}) \text{ with} \\
&\boldsymbol{iv}_{\text{vot01}} = (\text{seq01 seq02})^\top \text{ and } \boldsymbol{vs}_{\text{vot01}} = (\mathcal{VS}_1\ \mathcal{VS}_2\ \mathcal{VS}_3)^\top, \text{ where} \\
&\mathcal{VS}_1 = \left\{(1\,*)^\top, (*\,1)^\top\right\}, \mathcal{VS}_2 = \mathcal{VS}_3 = \left\{(*\,*)^\top\right\}.
\end{aligned}$$

The final element part channel consists of io-card, barrier, solenoid and valve. The examples will outline different possible situations in the lifecycle of the considered SIS. The

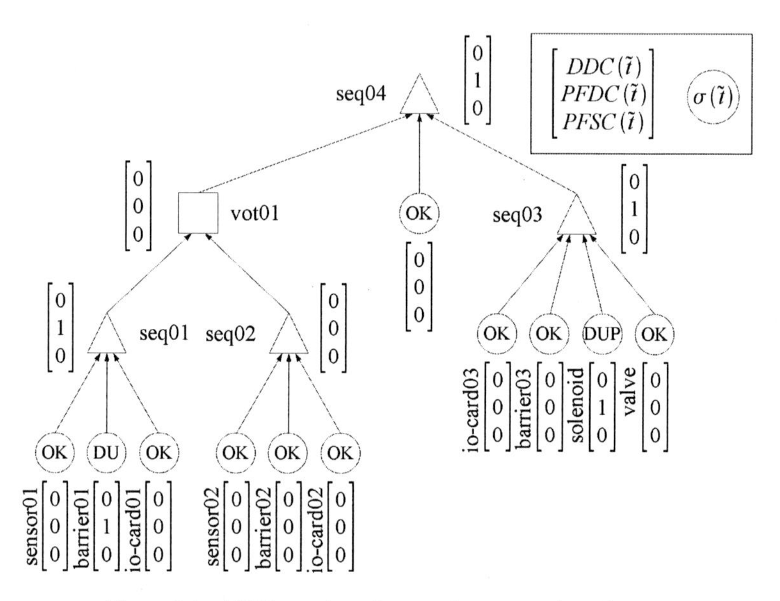

Figure 5.9: ASML graph evaluation for ex. 5.5 (p. 93)

related ASMLG is denoted in fig. 5.4 (p. 79). In this first example, the component state configuration vector $\tilde{\boldsymbol{\sigma}}$ is chosen as

$$\tilde{\boldsymbol{\sigma}} := (\sigma_{\text{sensor01}}(\tilde{t})\ \sigma_{\text{barrier01}}(\tilde{t})\ \sigma_{\text{io-card01}}(\tilde{t})\ \sigma_{\text{sensor02}}(\tilde{t})\ \sigma_{\text{barrier02}}(\tilde{t})\ \ldots$$
$$\ldots\ \sigma_{\text{io-card02}}(\tilde{t})\ \sigma_{\text{SPLC}}(\tilde{t})\ \sigma_{\text{io-card03}}(\tilde{t})\ \sigma_{\text{barrier03}}(\tilde{t})\ \sigma_{\text{solenoid}}(\tilde{t})\ \sigma_{\text{valve}}(\tilde{t}))^{\top}.$$
$$\tilde{\boldsymbol{\sigma}} = (OK\ DU\ OK\ OK\ OK\ OK\ OK\ OK\ OK\ DUP\ OK)^{\top}.$$

The results of the static ASMLG evaluation is depicted in fig. 5.9. From $\tilde{\boldsymbol{\sigma}}$ it can be derived that the barrier in the first sensor channel suffers from a dangerous undetected (*du*) failure and that the solenoid in the final element part is in its *DUP* failure state, i.e., contains a failure revealable by a partial proof test such as a partial valve stroke test. As no dangerous detected failures are in the SIS, all DDC functions carry a zero.

Sequence seq01 is s-unavailable (PFDC(seq01, $\tilde{t}$) $=$ *true*), as the second input of the sequence has failed dangerously (PFDC($\boldsymbol{iv}_{\text{seq01}}[2], \tilde{t}$) $=$ PFDC(barrier01, $\tilde{t}$) $=$ *true*), and the subsequent input, io-card01, is in its *OK* state. Since all components in the second sensor channel are in *OK* state, the superordinated sequence seq02 contributes to neither PFD nor PFS.

Voter vot01 operates on seq01 and seq02. The standard 1oo2 voting scheme is active, as all *dd* carrier functions are zero, i.e., DDC(seq01, $\tilde{t}$) $=$ DDC(seq02, $\tilde{t}$) $= 0$. A shutdown combination vector $(*\ 1) \in \mathcal{VS}_{\text{vot01},cur}$ can be found which has a one-entry for every index that represents a voter input j with $\text{PFDC}(\boldsymbol{iv}_{\text{vot01}}[j], \tilde{t}) = false$. Here, it is the second sensor channel with index $j = 2$. Hence, the voter is s- as well as o-available.

Sequence seq03 can easily be evaluated as PFD contributing, since its input with index 3 (the solenoid) evaluates as $\text{PFDC}(\boldsymbol{iv}_{\text{seq03}}[3], \tilde{t}) = \text{PFDC}(\text{solenoid}, \tilde{t}) = true$. And since the subsequently allocated valve does not contain an active failure, the overall final element channel seq03 has to be considered s-unavailable as well.

Finally, the overall unavailability state of the SIS can be classified as PFD contributing. The system output is represented by the evaluation results of sequence seq04. As the last related input vertex (seq03) contributes to s-unavailability, $\text{PFDC}(v_{SIS}, \tilde{t}) = \text{PFDC}(\text{seq04}, \tilde{t}) = true$.

Thus, the considered SIS at time $\tilde{t}$ is s-unavailable, i.e., would not respond properly in case of a demand. Since not a single active failure is contained in the system, the SIS can easily be identified as o-available.

A more sophisticated example will refer to a component state configuration of higher complexity.

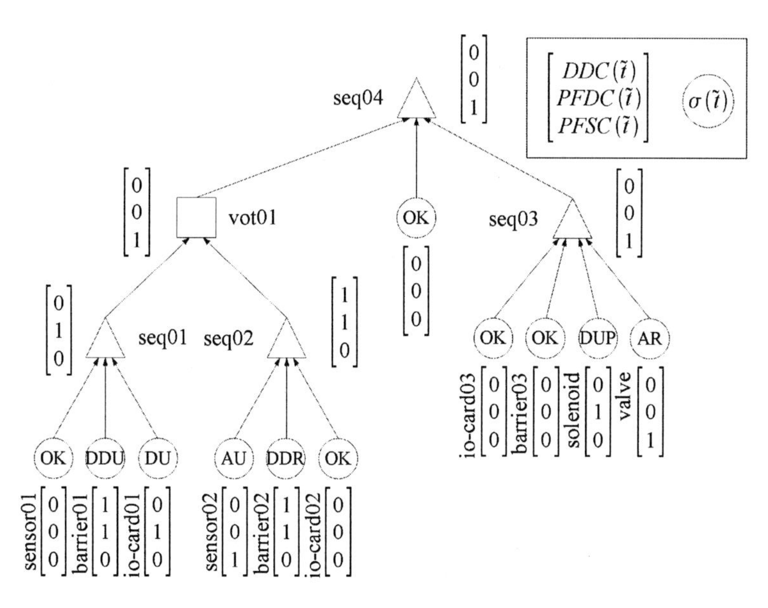

Figure 5.10: ASML graph evaluation for ex. 5.6 (p. 95)

Example 5.6 (Component state configuration 2 for SIS from fig. 2.1 (p. 9)). For this second example, the components state configuration vector $\tilde{\boldsymbol{\sigma}}$ is set up as follows:

$$\tilde{\boldsymbol{\sigma}} = (OK\ DDU\ DU\ AU\ DDR\ OK\ OK\ OK\ OK\ DUP\ AR)^{\top},$$

as shown in fig. 5.10. From $\tilde{\boldsymbol{\sigma}}$ it can be derived that barrier01 in the first sensor channel has failed with a dangerous detected failure. At the point of the occurrence of the failure there must already have been a dangerous undetected failure at the io-card01 such that the *dd* failure information did not immediately reach the SPLC and the component thus entered its DDU state. barrier02 in the second sensor channel failed with a *dd* failure which got immediately announced and led to the initiation of a repair process. Therefore, barrier02 is in DDR state. While this component has been under repair, an active (a) failure in sensor02 occurred. As the subsequently located and currently repaired *dd* failure inhibited this new failure, the sensor entered AU state, indicating that the failure has not been announced yet. The solenoid in the final element part of the SIS is in DUP state and the valve failed with an active (a) failure which is now under repair (AR).

Sequence seq01, representing the first sensor channel, evaluates as $\text{PFDC}(\text{seq01}, \tilde{t}) = \mathit{true}$, i.e., s-unavailable. This derives from seq01's third input vertex (io-card01) which is PFD contributing due to the DU state of the considered component. Notice that the sequence has a *dd* carrier value of $\text{DDC}(\text{seq01}, \tilde{t}) = 0$, since the *dd* failure at barrier01 is inhibited by the passive failure at io-card01.

The active failure at sensor02 in the second channel is inhibited and therefore not capable of impacting on seq02's o-availability, i.e., $\text{PFSC}(\text{seq02}, \tilde{t}) = \mathit{false}$. Instead, barrier02 in DDR state dictates the sequence's unavailability, as the very last input of seq02, i.e., io-card02 contains no failure. Therefore $\text{PFDC}(\text{seq02}, \tilde{t}) = \mathit{true}$ and additionally $\text{DDC}(\text{seq02}, \tilde{t}) = \text{DDC}(\text{barrier02}, \tilde{t}) = 1$, since one detected failure is carried by the sequence.

This *dd* failure reaches the voter vot01 and causes a degrading by rendering $\boldsymbol{vs}_{\text{vot01}}[2] = \left\{(*\,*)^{\mathsf{T}}\right\}$ the current voting scheme at time $\tilde{t}$. Since this shutdown combination vector represents a forced shutdown, voter01 evaluates as $\text{PFSC}(\text{vot01}, \tilde{t}) = \mathit{true}$. Notice that the voter causes a false trip of the SIS without having an active failure being the reason for it. In this case the chosen degradation scheme led to the regarded situation.

The unavailability state of seq03 depends on its right most input with index 4 which contributes to the PFS: $\text{PFSC}(\text{seq03}, \tilde{t}) = \text{PFSC}(\boldsymbol{iv}_{\text{seq03}}[4], \tilde{t}) = \text{PFSC}(\text{valve}, \tilde{t}) = \mathit{true}$.

The SIS's output vertex evaluates as $\text{PFSC}(v_{SIS}, \tilde{t}) = \text{PFSC}(\text{seq04}, \tilde{t}) = \mathit{true}$, since the last most input vertex of seq04 is o-unavailable: $\text{PFSC}(\boldsymbol{iv}_{\text{seq04}}[3], \tilde{t}) = \text{PFSC}(\text{seq03}, \tilde{t}) = \mathit{true}$.

The SIS is in spurious trip state for two reasons. Firstly, the DDR failure at barrier02 leads to a degradation of vot01 which corresponds to a forced shutdown. Secondly, the valve's active failure in the single final element part channel enforces an overall safe state. Notice that the active failure in the final element part alone would have been sufficient to cause the SIS spurious trip.

Example 5.7 (Component state configuration for sensor part with signal replication)**.** The sensor part given as extended ASMLG in fig. 5.11 consists of one sensor element (sensor) feeding two different sensor channels consisting of a barrier each (barrier01 and barrier02). This means that the sensor signal is splitted and injected twice into the SIS. Both (now interdependent) channels are connected to vot01 which performs a standard

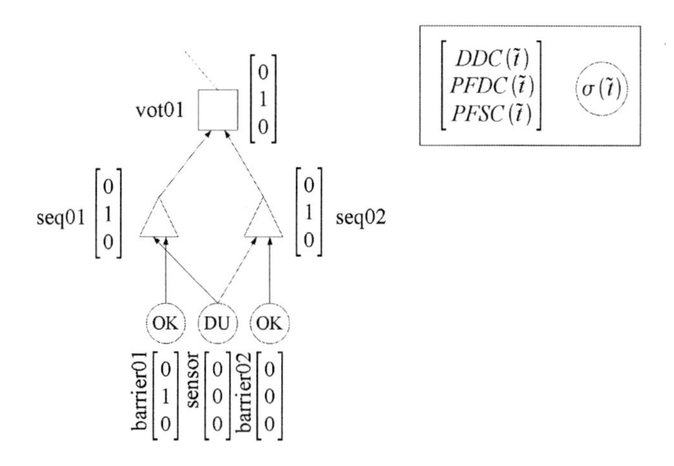

Figure 5.11: ASML graph with signal replication

static 1oo2 voting as shown in the previous examples. In contrast to the problems outlined in the paragraph about signal replication in subsec. 4.6.1 (p. 66) - R2, the ASML approach enables for a correct system evaluation in case of replicated signals. The exemplary SIS provided in fig. 5.11 is very similar to the one given as FT in fig. 4.23 (p. 67).

An exemplary component state configuration vector $\tilde{\boldsymbol{\sigma}}$ is set up as follows:

$$\tilde{\boldsymbol{\sigma}} := \left(\sigma_{\text{barrier01}}(\tilde{t})\ \sigma_{\text{sensor}}(\tilde{t})\ \sigma_{\text{barrier02}}(\tilde{t})\right)^{\top}$$
$$\tilde{\boldsymbol{\sigma}} = (OK\ DU\ OK)^{\top}.$$

The sensor element evaluating as PFDC(sensor, $\tilde{t}$) $= \mathit{true}$ serves as a basis for the subsequent considerations.

Both sequences' unavailability state depends on the replicated sensor signal, as both subsequently allocated barrier components are in their respective OK state. Hence, PFDC(seq02, $\tilde{t}$) = PFDC(seq02, $\tilde{t}$) $= \mathit{true}$ holds.

The voter vot01 has both input vertices s-unavailable. Since in the current 1oo2 voting scheme $\mathcal{VS}_{cur} = \left\{(1\ *)^{\top}, (*\ 1)^{\top}\right\}$ no suitable shutdown combination vector can be found, PFDC(vot01, $\tilde{t}$) $= \mathit{true}$.

As both sensor channels in the exemplary SIS rely on the functionality of a single sensor element, the overall system part is necessarily s- unavailable in case of a safety relevant failure at said device.

5.4 ASML SIS

In order to describe a SIS entirely, some additional specification is required. Up to now, e.g., the aforementioned maintenance groups (R7) have not been included in the ASML description language. The subsequent formal description of a SIS will complete the descriptive part of the approach outlined in this work.

Definition 5.22 (ASMLSIS). *An ASMLSIS is the basic mathematical structure, holding all required information about the safety instrumented function to be formally described. It is a four-tuple according to*

$$\mathcal{ASMLSIS}\left(\tau_{mt}, \mathcal{MGS}, \underbrace{\mathcal{V}, \mathcal{E}}_{ASMLG}\right),$$

where τ_{mt} is a mission time, $\mathcal{MGS}$ is a set of maintenance groups (see def. 5.23), $\mathcal{V}$ are the attributized vertices, and $\mathcal{E} \subset (\mathcal{V} \times \mathcal{V})$ the edges of the related ASMLG (see def. 5.3 (p. 78)).

The invention of a mission time for the SIS is a direct tribute to requirements R5, R6, and R9. Since all maintenance intervals for (partial) proof tests and component mission time replacements are to be specified for each component individually, the resulting system unavailability will typically not be periodic as, e.g., shown in subsec. 3.5.3. The SIS's mission time therefore defines the time span over which the required averaging process is performed (the PFD_{avg} is an averaged measure, see eq. 3.8 (p. 35)). τ_{mt} therefore specifies the time from commissioning until decommissioning.

The set of maintenance groups $\mathcal{MGS}$ is a specification for requirement R7:

Definition 5.23 (Maintenance group). *An arbitrary maintenance group $\mathcal{MG}_i$ in the finite set of maintenance groups $\mathcal{MGS}$ is a subset of the set of components*

$$\mathcal{MGS} = \{\mathcal{MG}_1, \mathcal{MG}_2, \ldots\} \text{ with } \mathcal{MG}_i \subseteq \mathcal{COM}$$

Whenever maintenance is applied to any component $c_m \in \mathcal{MG}_i$, a functional test which also impacts on all $c_n \in \mathcal{MG}_i$ with $c_n \neq c_m$ is performed. Components may be assigned to more than one maintenance group at once:

$$|\mathcal{MG}_1 \cap \mathcal{MG}_2 \cap \ldots| \geq 0.$$

If an arbitrary component undergoes a proof test, then all components assigned to maintenance groups with the considered component as member receive a proof test, too. Subsequent parts of this work require to find all of these affected components. This can be accomplished with the maintenance group function:

Definition 5.24 (Maintenance group function). *The maintenance group function*

$$\begin{aligned}
&\text{MG} : \mathcal{COM} \mapsto \mathcal{COM} \times \mathcal{COM} \times \ldots := \\
&\text{MG}(c_m) = \{c_n | \, (c_n \neq c_m) \wedge (\exists (\mathcal{MG}_i \in \mathcal{MGS}) \, (\{c_m, c_n\} \subseteq \mathcal{MG}_i))\}
\end{aligned}$$

delivers a set that consists of all components from all maintenance groups, not containing the component c_m.

A short example will close this subsection:

Example 5.8 (Exemplary ASMLSIS parameterization). The SIS denoted in fig. 2.1 (p. 9) with related ASMLG according to fig. 5.3 (p. 78) could be parameterized as shown in table 5.2. It is specified as a SIS with a mission time of ten years, and maintenance groups each including all components of the respective channels in sensor part and final element part. The choice of the maintenance groups induces that sensor channels are tested using a function test button and final elements via SPLC trigger signal as described in sec. 2.2 (p. 10). The maintenance group function MG delivers (exemplarily)

$$\begin{aligned} \mathrm{MG}(\mathrm{sensor01}) &= \{\mathrm{barrier01}, \mathrm{io\text{-}card01}\}\,, \\ \mathrm{MG}(\mathrm{solenoid}) &= \{\mathrm{io\text{-}card03}, \mathrm{barrier03}, \mathrm{valve}\}\,. \end{aligned}$$

Table 5.2: Example parameterization for the ASMLSIS denoted in fig. 2.1 (p. 9) (times in h)

Parameter	Value(s)
$\mathcal{T}_{mt}$	87600
$\mathcal{MGS}$	$\{\mathrm{sensor01}, \mathrm{barrier01}, \mathrm{io\text{-}card01}\}$, $\{\mathrm{sensor02}, \mathrm{barrier02}, \mathrm{io\text{-}card02}\}$, $\{\mathrm{io\text{-}card03}, \mathrm{barrier03}, \mathrm{solenoid}, \mathrm{valve}\}$
$\mathcal{V}$	$\mathcal{V}_{fig.\ 5.3}$
$\mathcal{E}$	$\mathcal{E}_{fig.\ 5.3}$

5.5 Practical ASML description guidelines

As ASML aims at generalizing many relevant effects for SIS in the process industry, it appears difficult to express standard problems. This section intends to provide a short overview on how to model/engineer certain effects such as mechanical components or common cause scenarios (see fig. 5.12). Thereby transformation formulas are introduced that take commonly utilized parameterization as input and generate valid ASML descriptions automatically.

5.5.1 MooN voting schemes

Although MooN voting schemes are practically standard in modern safety loops, non-standard voting can be found more and more often. The ASML voting scheme approach as introduced in def. 5.15 and def. 5.16 (p. 87) provides lots more flexibility while additionally allowing for the description of dynamic voting, i.e., degradations. On the other hand it might be difficult to identify a given voting scheme as, e.g., a 2oo3 redundancy, by looking at the shutdown combination vectors. Mathematically it is easy to generate the appropriate set of shutdown combination vectors for any given MooN voting:

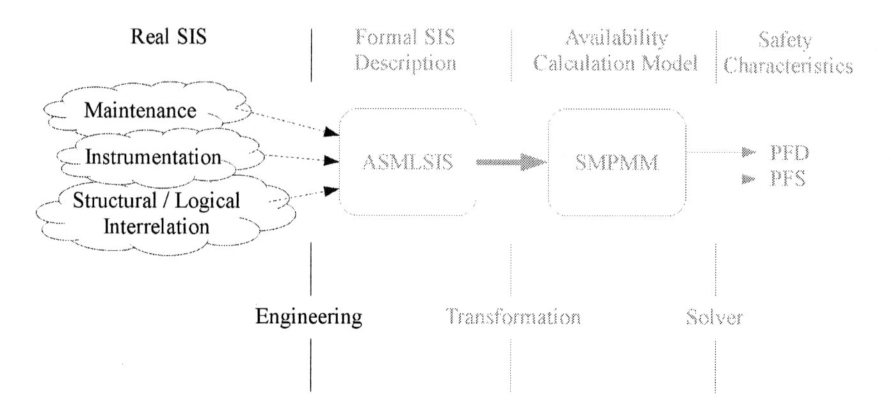

Figure 5.12: Flow diagram of the ASML approach and location of the description guidelines

Definition 5.25 (Voting scheme generation for MooN specifications). *The voting scheme for M-out-of-N redundancies is a set of $\binom{N}{M}$ shutdown combinations as introduced in def. 5.15 (p. 86) with special properties:*

$$\mathcal{VS}_{MooN} = \left\{ \boldsymbol{sdc} \middle| \sum_{i=1}^{N} \boldsymbol{sdc}[i] = M \right\}.$$

As ASML descriptions are typically edited via a graphical user interface (an editor), a reasonable approach is to allow the user to specify an arbitrary MooN voting and generate the appropriate shutdown combination vectors via def. 5.25 which algorithmically leads to a simple iteration problem.

5.5.2 Periodic interaction with components

Often the available flexibility for conducting proof tests or replacing components etc. is not exploited entirely. Usually a constant proof test interval is utilized along with a static partial proof test and a replacement policy. The construction of the related required ASML sets from a constant time interval can be conducted easily:

Definition 5.26 (Periodic component interaction). *The proof test set $\mathcal{PT}_i$, the partial proof test set $\mathcal{PPT}_i$ and the mission time set $\mathcal{MT}_i$ for an arbitrary component $c_i \in \mathcal{COM}$ can be constructed from constant intervals $\tau_{pt,i}$, $\tau_{ppt,i}$ and $\tau_{mt,i}$ according to*

$$\mathcal{PT}_i = \{t \in \mathbb{R}_+ \mid (t \ mod \ \tau_{pt,i} = 0) \wedge (t < \tau_{mt})\},$$

$$\mathcal{PPT}_i = \{t \in \mathbb{R}_+ \mid (t \ mod \ \tau_{ppt,i} = 0) \wedge (t < \tau_{mt})\},$$

$$\mathcal{MT}_i = \{t \in \mathbb{R}_+ \mid (t \ mod \ \tau_{mt,i} = 0) \wedge (t < \tau_{mt})\} ,$$

where τ_{mt} *is the mission time of the SIS.*

5.5.3 Mechanical components

DIN EN 61511 refers to mathematic approaches as proposed by DIN EN 61508 for PFD calculation purposes. Therefore, the exponential life model as introduced in sec. 3.3 (p. 32) is indirectly assumed for each type of component. Thus, the ASML approach demands constant failure rates. This leads to problems when dealing with mechanical equipment. In contrast to electronic devices, the occurring wear out effects cannot be described using an exponential life distribution. Instead, the Weibull distribution gives an appropriate failure model for mechanical components as long as the component is operated in its wear out effect range. Notice that below a certain switching frequency other effects due to disuse become relevant. These effects cannot be described with the Weibull distribution. The discrete CDF for mechanical component in wear out mode is given as

$$\mathrm{P}(CTF \leq c) = \mathrm{F_m}(c) = 1 - \exp\left[-\left(\frac{c}{C}\right)^b\right],$$

where c is the number of load cycles for the considered component, e.g., a relay, and CTF the random variable 'cycles-to-failure'. C is the scale parameter or characteristic lifetime and b the shape parameter or Weibull modul. In order to do PFD calculations utilizing mechanical components an exponential approximation of the Weibull distribution has to be applied. This is accomplished by introducing the B10 value [SN 07]. It determines the number of load cycles after which 10% of a given population of equal mechanical components have statistically failed. In the area of mechanical engineering it is good engineering practice to replace mechanical components after B10 load cycles by maintenance directive. If this principle is applied to mechanical safety equipment, a maximum useful lifetime can be determined as

$$\tau_{mt,m} = \mathrm{B10} \, \frac{8760}{h_{op} d_{op} f_{cycle}}, \tag{5.1}$$

where h_{op} is the number of operational hours per day, d_{op} is the number of operational days per year and f_{cycle} the switching frequency in Hz. Equation 5.1 transforms the absolute number of load cycles B10 into the time domain. For c = B10 the related discrete Weibull distribution thus obviously delivers $\mathrm{F_m}(c = \mathrm{B10}) = 0.1$ due to the definition of the B10 value. The required approximation can now be derived by finding the exponential distribution $\overline{\mathrm{F}}_m(t)$ which delivers an unreliability of 0.1 at the end of the useful lifetime

$\tau_{mt,m}$. It can be determined by setting

$$\begin{aligned}\overline{\mathrm{F}}_m(t=\tau_{mt,m}) &= \mathrm{F}_\mathrm{m}(c=B_{10})\\ \Rightarrow 1-\exp(-\lambda_m\tau_{mt,m}) &= 0.1\\ \lambda_m &= -\frac{\ln 0.9}{\tau_{mt,m}}\\ &\approx \frac{0.1}{\tau_{mt,m}}.\end{aligned}$$

The relevant λ_m to be used with the ASML approach therefore is (using 5.1):

$$\lambda_m = \frac{0.1}{\mathrm{B10}}\frac{h_{op}d_{op}f_{cycle}}{8760}.$$

Similar equations can be found, e.g., in [DIN07], as well as in [SN 07]. Thus, PFD calculation with mechanical components is admissible if

- the approximation is done correctly
- the mechanical device gets replaced after $t=\tau_{mt,m}$ or B10 load cycles, respectively
- the switching frequency and the duty cycle do not cause ageing effects that cannot be described by the Weibull distribution
- the B10 value is valid for the operational switching frequency
- the device is operated under the conditions defined by the manufacturer

5.5.4 Beta model reproduction

An easy way of reproducing effects of the beta model can be derived from the common method of expressing CC failures as individual primary events in FTs [IEC09]. This way, they behave as if they were individual separate components that are connected with the non-CC part via a 2oo2 voting. Consider fig. 5.13: the ASMLG on the left depicts a 2oo3 voting SIS. Assuming homogeneous instrumentation ($\lambda_{com01,du}=\lambda_{com02,du}=\lambda_{com03,du}=\lambda_{du}$ and $\lambda_{com01,dd}=\lambda_{com02,dd}=\lambda_{com03,dd}=\lambda_{dd}$ according to IEC 61508), a 'virtual' component com04' can be constructed (on the right in fig. 5.13), that represents the CC failure fraction. Its *du* failure rate can be calculated as $\lambda_{com04',du}=\beta\lambda_{du}$. The failure rate related to *dd* failures is $\lambda_{com04',dd}=\beta_D\lambda_{dd}$. As the CC failure fraction has now been modeled as a separate component, the related fraction of the failure rate has to be removed from the original components. The modified components com01', com02', and com03' have failure rates according to $\lambda_{com01',du}=\lambda_{com02',du}=\lambda_{com03',du}=(1-\beta)\lambda_{du}$ and $\lambda_{com01',dd}=\lambda_{com02',dd}=\lambda_{com03',dd}=(1-\beta_D)\lambda_{dd}$. The CC component com04' is connected to the new 2oo2 voter vot02 together with the output of vot01. The original 2oo3 voter vot01 has been taken over to the ASMLG on the right of fig. 5.13 as it is.

The outlined method of reproducing effects of the beta model is somewhat of a workaround, as more effective ways exist of explicitly implementing CC failures into the

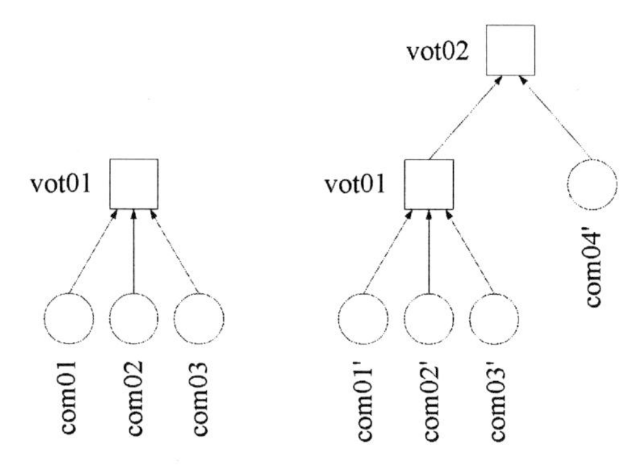

Figure 5.13: Reproduction of the beta model in a 2oo3 redundant system

MM using modified transitions (see, e.g., [Gob00]). But as discussed in subsec. 2.3.6 (p. 21), resulting in R11, an explicit implementation is not the aim of this work. The outlook in chapter 9 deals with further contributions to the ASML approach that could deal with the inclusion of a more general CC failure model that might include the beta model as a special case.

Chapter 6

Transformation: ASML description to SMPMM

6.1 Overview

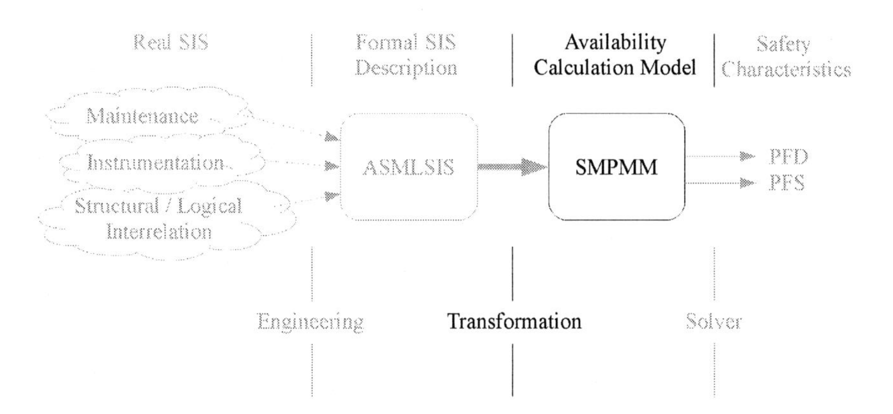

Figure 6.1: Flow diagram of the ASML approach and location of the SMPMM as well as the related transformation

Based on the ASML description of a safety loop, i.e., an ASMLSIS, a markov based calculation model can be automatically constructed. This is accomplished by introducing suitable generator formulas which are - together with the ASML description language from chapter 5 - the main contribution of this work. Figure 6.1 denotes the location of the transformation within the entire ASML approach.

This chapter firstly introduces safety multiphase markov models (SMPMM) as a new and generalized extension to standard multiphase markov models as introduced in subsec. 4.4.5 (p. 55). Along with the model, a set of solver equations is delivered (sec. 6.3

(p. 107)) enabling for retrieving the desired availability characteristics PFD and PFS.

Section 6.3 finally provides the set of generator formulas. SMPMM construction requires generating the state space (subsec. 6.3.1 (p. 107)), selection vectors (subsec. 6.3.2 (p. 109)), transition matrix (subsec. 6.3.3 (p. 109)), and phase transition matrices (subsec. 6.3.4 (p. 116)).

6.2 Safety multiphase markov model

The safety multiphase markov model is based on the MPMM as introduced in def. 4.1 (p. 56) with all extensions required for implementing the various requirements from sec. 2.4 (p. 24).

Definition 6.1 (Safety multiphase markov model). *A safety multiphase markov model (SMPMM) is a 7-tuple*

$$\mathcal{SMPMM}\left(\mathcal{S}, \boldsymbol{A}, \boldsymbol{p}_0, \mathcal{T}, k_{mt}, \boldsymbol{c}_{PFD}^{\top}, \boldsymbol{c}_{PFS}^{\top}\right),$$

where $\mathcal{S}$ is a state space of size Q, $\boldsymbol{A}$ is a $Q \times Q$ transition matrix. $\boldsymbol{p}_0$ is an initial probability distribution vector of dimension $Q \times 1$. $\mathcal{T}$ is a set of phase transitions containing tuples $(k_t, \boldsymbol{M}_t)$, where $k_t \Delta t$ are discrete points in time a phase transition is performed, and $\boldsymbol{M}_t$ are $Q \times Q$ phase transition matrices. $k_{mt} \Delta t$ is a discrete mission time. $\boldsymbol{c}_{PFD}^{\top}$ and $\boldsymbol{c}_{PFS}^{\top}$ are selection vectors of dimension $1 \times Q$.

Most important difference to the simple MPMM from def. 4.1 is the set of phase transitions $\mathcal{T}$. Instead of utilizing a single constant phase duration, it is necessary to provide a phase transition for each specified element in each component's sets $\mathcal{PT}$, $\mathcal{PPT}$, and $\mathcal{MT}$ (see def. 5.11 (p. 83)). Since all components are individually parameterized, it is easily possible that a single component changes its state at a phase transition or that large groups of components are, e.g., proof tested together. The shifting matrices $\boldsymbol{M}_t$ need to get constructed appropriately. The mission time for the SMPMM specifies the time span over which the averaging process for the unavailability characteristics is performed. Selection vectors need to be constructed for each of the relevant characteristics PFD and PFS. Notice that all information is derived from one single markov model. As active and passive failures interact with each other as discussed in, e.g., sec. 2.4 (p. 24), independent calculation of PFD and PFS with decoupled markov models is not possible.

The SMPMM's dynamics can be described by the discrete time phase transition function:

Definition 6.2 (Phase transition function). *The discrete time phase transition function ϕ is defined as*

$$\phi(\mathcal{T}, \boldsymbol{A}, k) = \begin{cases} \boldsymbol{M}_t \boldsymbol{A} & \text{if } \exists\, ((k_t, \boldsymbol{M}_t) \in \mathcal{T})\, (k_t = k+1) \\ \boldsymbol{A} & \text{else.} \end{cases}$$

The phase transition function ensures that for each discrete time step a possible phase transition is detected and - if required - a probability shifting process among the SMPMM is performed. The system's dynamics can thus be expressed by an extended recursive Chapman-Kolmogorov equation:

Definition 6.3 (SMPMM dynamics). *The discrete time probability distribution of an SMPMM can be obtained by utilizing subsequent recursive equation:*

$$\boldsymbol{p}(k+1) = \phi(\mathcal{T}, \boldsymbol{A}, k)\boldsymbol{p}(k) \text{ with } \boldsymbol{p}(0) = \boldsymbol{p}_0.$$

With the dynamics equation, the behavior of the unavailability characteristics over time can be retrieved.

Definition 6.4 (Unavailability over discrete time). *The discrete time unavailabilities of a SIS represented by an SMPMM can be calculated via*

$$\begin{aligned} \mathrm{PFD}(k) &= \boldsymbol{c}_{PFD}^{\top}\boldsymbol{p}(k), \\ \mathrm{PFS}(k) &= \boldsymbol{c}_{PFS}^{\top}\boldsymbol{p}(k). \end{aligned}$$

Based on the SIS's unavailability over time, the relevant averaged values can be obtained using simple arithmetic means.

Definition 6.5 (Averaged unavailability). *The averaged unavailabilities over k_{mt} discrete mission time steps of a SIS represented by an SMPMM can be obtained as*

$$\begin{aligned} PFD_{avg} &= \frac{1}{k_{mt}} \sum_{k=0}^{k_{mt}-1} \mathrm{PFD}(k), \\ PFS_{avg} &= \frac{1}{k_{mt}} \sum_{k=0}^{k_{mt}-1} \mathrm{PFS}(k). \end{aligned}$$

6.3 Generator formulas

6.3.1 State space

The state space of all MMs grows exponentially with the number of represented components. Therefore, the subset $\mathcal{COM}$ of the vertices $\mathcal{V}$ of an ASMLG plays an important role during the state space construction. The initial (i.e., maximum) number of states in a SMPMM can be obtained according to the subsequent definition:

Definition 6.6 (Size of the state space).

$$Q = |\mathcal{CS}|^{|\mathcal{COM}|}.$$

is the number of state space vectors in an SMPMM's state space $\mathcal{S}$.

The underlying combinatorial problem is: choose $|\mathcal{COM}|$ component states out of $|\mathcal{CS}|$ available states. Multiple choice of states is possible, the order of the chosen states is relevant [Har09]. For reasons of compact notation, the number of components $|\mathcal{COM}|$ will be abbreviated:

Definition 6.7 (Number of components).

$$N_{SIS} = |\mathcal{COM}|\,.$$

is the number of components in an ASML SIS.

Utilizing these two definitions as well as the definition for the component's state space according to def. 5.1 (p. 72), the state space can be defined:

Definition 6.8 (State space generation). *Let*

$$\mathcal{S} = \{\boldsymbol{s}_1, \ldots, \boldsymbol{s}_Q\}$$
$$\textit{with } \boldsymbol{s}_i \in \left\{(\boldsymbol{s}[1] \ \ldots \ \boldsymbol{s}[N_{SIS}])^\top \mid (\boldsymbol{s}[1], \ldots, \boldsymbol{s}[N_{SIS}]) \in (\mathcal{CS} \times \ldots \times \mathcal{CS})\right\}$$

be the state space of the SMPMM, i.e., a set containing Q vectors $\boldsymbol{s}_i$ of dimension N_{SIS} and $\boldsymbol{s}_i[j] \in \mathcal{CS}$. Each vector in the state space can be considered a component state configuration vector $\tilde{\boldsymbol{\sigma}}$ according to def. 5.21 (p. 93). The order of the vectors' elements corresponds to the order of the components in ASMLGs from left to right (if not defined otherwise).

As the component state set $\mathcal{CS}$ contains all possible output values for the component state functions $\sigma_i(t)$ (compare def. 5.12 (p. 84)), the state space $\mathcal{S}$ reflects *all* possible permutations of $\sigma_i(t)$ outputs, i.e., *all* potential overall system states.

The basic idea of the SMPMM's state space shall be clarified by a simple example.

Example 6.1 (State space generation for a two component system). With the component state set $\mathcal{CS}$ as defined in def. 5.1 (p. 72) and $N_{SIS} = 2$ components, the resulting state space has $Q = |\mathcal{CS}|^{N_{SIS}} = 9^2 = 81$ elements. Subsequently an excerpt from the state space:

$$(OK\,OK)^\top, (OK\ AU)^\top, (OK\ AR)^\top, (OK\ DDU)^\top, (OK\ DU)^\top,$$
$$(OK\ DUP)^\top, (OK\ DN)^\top, (OK\ DDR)^\top, (OK\ DR)^\top, (AU\,OK)^\top, \ldots$$
$$\ldots, (DR\,OK)^\top, (DR\ AU)^\top, (DR\ AR)^\top, (DR\ DDU)^\top, (DR\ DU)^\top,$$
$$(DR\ DUP)^\top, (DR\ DN)^\top, (DR\ DDR)^\top, (DR\ DR)^\top$$

It is good engineering practice to assume an SIS starting in a state where all components are in their respective functional states.

Definition 6.9 (Initial probability distribution). *The initial probability distribution for an SMPMM is (if not explicitly specified differently) assumed as*

$$\boldsymbol{p}_0 = (p_1\, p_2\, p_3\, \cdots\, p_Q)^\top = (1\,0\,0\, \cdots\, 0)^\top, \textit{ where } p_1 = \mathrm{P}(\boldsymbol{s} = \boldsymbol{s}_1 = (OK\ OK\ \cdots\ OK)^\top).$$

6.3.2 Selection vectors

In def. 5.21 (p. 93), the static ASMLG evaluation has been introduced. It allows for the determination of the SIS's safety related as well as operational unavailability for a given assignment of the component state functions $\sigma_i(t)$ at $t = \tilde{t}$. Selection vectors as introduced in subsec. 4.4.1 (p. 50) are required to pick all state probabilities from a MM that contribute to the desired unavailability characteristic. The following definitions for the generation of selection vectors demonstrate how the static ASMLG evaluation functions are predestined to support the generic construction of SMPMM.

Definition 6.10 (Selection vector generation). *The selection vectors of size $1 \times Q$ for unavailability measures PFD and PFS specify the respective relevant states in the SMPMM's state space:*

$$\boldsymbol{c}_{PFD}[i] = \begin{cases} 1 & \textit{if } \mathrm{PFDC}(v_{SIS}, \tilde{t}_i) = 1, \\ 0 & \textit{else.} \end{cases}$$

$$\boldsymbol{c}_{PFS}[i] = \begin{cases} 1 & \textit{if } \mathrm{PFSC}(v_{SIS}, \tilde{t}_i) = 1, \\ 0 & \textit{else.} \end{cases}$$

$\tilde{t}_i$ is related to a component state configuration vector $\tilde{\boldsymbol{\sigma}}_i$, such that $\tilde{\boldsymbol{\sigma}}_i = \boldsymbol{s}_i$ holds, i.e., a static ASMLG evaluation according to subsec. 5.3.7 (p. 92) is performed for each state $\boldsymbol{s}_i$ of the state space $\mathcal{S}$. $\boldsymbol{c}_{(\cdot)}[i] = 1$ reads as 'the probability of the i'th SMPMM state contributes to the desired unavailability characteristic.

An example shall clarify the process of shutdown vector generation.

Example 6.2 (Selection vector generation for the ASMLG from fig. 5.11). Reconsider the ASMLG from fig. 5.11 (p. 97) with static 1oo2 voting. Assuming, that the depicted graph represents the whole SIS, an excerpt from the related state space is

$$\begin{aligned} \boldsymbol{s}_1 &= (OK\ OK\ DU)^\top, \boldsymbol{s}_2 = (OK\ DU\ OK)^\top, \boldsymbol{s}_3 = (DU\ OK\ OK)^\top, \\ \boldsymbol{s}_4 &= (AR\ DU\ OK)^\top, \boldsymbol{s}_5 = (DU\ DDU\ OK)^\top, \boldsymbol{s}_6 = (DDR\ OK\ OK)^\top, \ldots, \end{aligned}$$

where $\tilde{\boldsymbol{\sigma}} := \left(\sigma_{\text{barrier01}}(\tilde{t})\ \sigma_{\text{sensor}}(\tilde{t})\ \sigma_{\text{barrier02}}(\tilde{t})\right)^\top$. The related selection vectors are generated as

$$\boldsymbol{c}_{PFD} = (0\,1\,0\,0\,1\,0\ \ldots)^\top, \boldsymbol{c}_{PFS} = (0\,0\,0\,1\,0\,0\ \ldots)^\top.$$

6.3.3 Transition matrix

Central object in the mathematical description of a SIS as SMPMM is the transition matrix which fulfills the important purpose of specifying a system's dynamics disregarding external intervention by, e.g., phase transitions. In contrast to all considerations made in sec. 4.4 (p. 50) about markov models, the transition matrix required for the generic approach needs to reproduce heavy interdependencies among the modeled components. According to the requirements from sec. 2.4 (p. 24), components may change their internal state if, e.g., a local inhibition vanishes or a component in the local maintenance group gets repaired. These events are shown and described in the CFSM from subsec. 5.2.3 (p. 73) as $\boldsymbol{e_{!inhibited}}$ and $\boldsymbol{e_{mgr}}$. The subsequent generator formulas construct a transition matrix under consideration of the mentioned effects. Naturally, all exponentially distributed failure and repair processes according to the CSFM events $\boldsymbol{e_{F,du}}$, $\boldsymbol{e_{F,dup}}$, $\boldsymbol{e_{F,dd}}$, $\boldsymbol{e_{F,ddx}}$, $\boldsymbol{e_{F,dn}}$, $\boldsymbol{e_{F,a}}$, $\boldsymbol{e_{R,ar}}$, $\boldsymbol{e_{R,dr}}$, $\boldsymbol{e_{R,ddr}}$ are implemented as well.

As the transition matrix generator functions are rather complex, it is useful to define some elementary functions that support the generator functions and allow for shorter notation.

The vector difference function delivers a set of indices. It is heavily used for comparing two state space vectors with the intention of investigating whether a markov transition between them is possible.

Definition 6.11 (Vector difference set function). *The vector difference set function* Diff *delivers a set of element indices for which two vectors* $\boldsymbol{a}$ *and* $\boldsymbol{b}$ *of equal size differ from each other regarding the elements' respective values.*

$$\begin{aligned} &\text{Diff} : \mathbb{R}^m \times \mathbb{R}^m \mapsto \{d_1, \dots, d_n\}, d_i \in \mathbb{N}_+, d_n \leq m \\ &:= \text{Diff}(\boldsymbol{a}, \boldsymbol{b}) = \{i \mid \boldsymbol{a}[i] \neq \boldsymbol{b}[i]\} \end{aligned} \tag{6.1}$$

Example 6.3 (Simple vector difference sets). A short example shall clarify the idea of vector difference sets:

$$\begin{aligned} &\text{Diff}((21\ 8\ 17\ 1\ 1)^\top, (1\ 8\ 17\ 4\ 1)^\top) = \{1, 4\}, \\ &\text{Diff}((1\ 2\ 3\ 4\ 5)^\top, (1\ 2\ 5\ 4\ 3)^\top) = \{3, 4, 5\}, \\ &\text{Diff}((120\ 5\ 2)^\top, (120\ 5\ 2)^\top) = \{\}, \end{aligned}$$

The inhibition function serves the purpose of verifying whether a certain component failure to occur shortly would be inhibited or not depending on the state of all other components. This function therefore implements the functionality of the state variable ***inhibited*** described in context with the CFSM from subsec. 5.2.3 (p. 73).

Definition 6.12 (Inhibition set function). *The inhibition set function* Inh *is defined as*

$$\begin{aligned} &\text{Inh} : \mathcal{S} \mapsto \{d_1, \dots, d_n\}, d_i \in \mathbb{N}, n \leq N_{SIS} \\ &:= \text{Inh}(\boldsymbol{s}_m) = \left\{i \mid \nexists\, (v_p \in \mathcal{VOT})\, \nexists q\, \left(\text{PFSC}(\boldsymbol{iv}_p[q], \tilde{t})\right) \wedge !\, \text{PFSC}(v_{SIS}, \tilde{t})\right\}, \end{aligned}$$

where

$$\tilde{t} \mid \forall j \left(\sigma_j(\tilde{t}) = \begin{cases} AR & \text{if } j = i, \\ DU & \text{if } (j \neq i) \wedge (\boldsymbol{s}_m[j] \neq OK), \\ OK & \text{else.} \end{cases} \right)$$

It delivers for a given vector $\boldsymbol{s}_m$ from the state space $\mathcal{S}$ a set with all component indices that belong to components with the following characteristic: if a dd or a failure instantly occurs for the considered component, this failure would be inhibited due to subsequently allocated failures.

Inhibition occurs in sequences only, and only if a component or subsystem allocated 'behind' an occurring *dd* or *a* failure is in a failure state. In other words: if a self revealable failure can be forwarded to a voter or the SIS output v_{SIS}, it is considered not inhibited.

For each potentially inhibited component with index i, a 'reference component state configuration vector' is constructed ($\tilde{t} = \cdots$). All active and passive failures in the original state space vector $\boldsymbol{s}_m$ are set to DU failure state. Only component i is set to AR. If the static evaluation with this reference vector delivers o-unavailability at any voter input or at the system output, then a potential active failure at i would reach an annunciation entity, i.e., would not be inhibited. This 'trick' is executed for each single component index.

Definition 6.12 can be expressed as an algorithm that enables for the investigation of this situation. This algorithm shall be outlined explicitly for clarifying the concept:

```
    Input: state space vector s_m
    Output: set of component indices IND
 1  IND = ∅;
 2  foreach component index i do
 3      foreach component index j do
 4          if j = i then σ_j(t̃) := AR;
 5          if j ≠ i and s_m[j] ≠ OK then σ_j(t̃) := DU;
 6          if j ≠ i and s_m[j] = OK then σ_j(t̃) := OK;
 7      end
 8      inhibition := true;
 9      foreach voter v_p do
10          foreach input vertex index q do
11              if PFSC(iv_p[q], t̃) = true then inhibition := false;
12          end
13      end
14      if PFSC(v_SIS, t̃) = true then inhibition := false;
15      if inhibition = true then IND := IND ∪ i;
16  end
```

Algorithm 6.1: Inhibition set function algorithm

The basic idea of algorithm 6.1 is as follows: if only one single active failure is present in a SIS and any of the voters' input vertices or the SIS output itself contributes to the PFS, then the said failure is obviously not inhibited. The algorithm implements this by firstly choosing a component index investigation candidate i (line 2). Secondly, the component state function for component i is set to AR, i.e., a failure state causing a PFS contribution at any voter or the system output if not inhibited (line 4). Thirdly, all other component state functions are set to DU if the related element of the input state space vector $\boldsymbol{s}_m$ is not OK. This ensures that no further active failures remain in the system (line 5). As a fourth step, all components without failure in $\boldsymbol{s}_m$ have their component state function set to OK (line 6). Afterwards, the algorithm cycles through all voters contained in the system and checks, whether any of them is PFS contributing (lines 9 - 13). If that is the case, the index under investigation i does definitely not represent a component that would have an instant dd or a failure inhibited. Hence, the index is not appended to the set of indices $\mathcal{IND}$ (line 15). The same situation is given, if the SIS output v_{SIS} (which needs not to be a voter) evaluates as $\text{PFSC}(v_{SIS}, \tilde{t}) = true$ (line 14). If the investigation is done for index i, the next investigation candidate is chosen for i. A simple example shall provide support in clarifying the working principle of the inhibition set function.

Example 6.4 (Simple inhibition set generation)**.** Reconsider fig. 5.10 (p. 95). The outlined component state configuration is interpreted as a state space vector $\boldsymbol{s} \in \mathcal{S}$:

$$\boldsymbol{s} = (OK\ DDU\ DU\ AU\ DDR\ OK\ OK\ OK\ OK\ DUP\ AR)^\top$$

The inhibition set for this state space vector is wanted ($\text{Inh}(\boldsymbol{s})$).

At first, the investigation candidate i is set to 1 (line 2). Then the component state configuration vector $\tilde{\boldsymbol{\sigma}}$ for the considered SIS is reconfigured, using lines 3 - 7:

$$\tilde{\boldsymbol{\sigma}} = (AR\ DU\ DU\ DU\ DU\ OK\ OK\ OK\ OK\ DU\ DU)^\top .$$

Afterwards, the ASMLG is statically evaluated. The single voter vot01 evaluates as $\text{PFSC}(\text{vot01}, \tilde{t}) = false$ (lines 9 - 13) and so does the system output: $\text{PFSC}(v_{SIS}, \tilde{t}) = false$ (line 14). Hence, $i = 1$ is contained in the inhibition set $\mathcal{IND}$ (line 15). This is intuitively clear, as a spurious trip of the sensor element in the first SIS channel would not reach the voter due to both subsequently allocated channel components (i.e., in the same sequence) being in passive failure states (DDU, DU).

Now the investigation candidate i is set to 2. Using lines 3 - 7 again, $\tilde{\boldsymbol{\sigma}}$ is configured as

$$\tilde{\boldsymbol{\sigma}} = (OK\ AR\ DU\ DU\ DU\ OK\ OK\ OK\ OK\ DU\ DU)^\top ,$$

resulting in the same consequence: $i = 2$ is contained in the inhibition set. This means that if barrier01 suddenly suffers a dd or a failure, it would not reach the voter get announced.

For $i = 3$ and

$$\tilde{\boldsymbol{\sigma}} = (OK\ DU\ AR\ DU\ DU\ OK\ OK\ OK\ OK\ DU\ DU)^\top ,$$

it can easily be retraced, that vot01 now evaluates as $\text{PFSC}(\text{vot01}, \tilde{t}) = true$, since no subsequent input vertex exists for sequence seq01. An instantly occurring dd or a failure at io-card01 would therefore be detected and announced immediately. Therefore, $i = 3$ is not contained in $\mathcal{IND}$.

After the completion of the iteration over all eleven component indices, the result of the inhibition set function is

$$\text{Inh}(\boldsymbol{s}) = \{1, 2, 4, 7, 8, 9, 10\} .$$

Property 6.1 (Single inhibition release principle). *Upon completion of an FR process for any component in a SIS, only a single by then inhibited failure might get released.*

With the vector difference set function and the inhibition set function, the transition matrix $\boldsymbol{A}$ can finally be constructed based on an ASMLSIS. The generator function works basically on the state space set $\mathcal{S}$ with indexed elements $\boldsymbol{s}_1, \boldsymbol{s}_2, \ldots$, the state space vectors. With an element $\boldsymbol{A}[m][n]$ the transition matrix encodes the probability for the transition from state $\boldsymbol{s}_n$ to state $\boldsymbol{s}_m$ in the time interval Δt. The basic concept is to iterate over all matrix elements of $\boldsymbol{A}$ and investigate the related state transition whether it is possible and - if yes - how the transition probability has to be chosen.

Definition 6.13 (Failure transition). *The transition matrix element $\boldsymbol{A}[m][n]$ with $\{m,n\} \in \{1,\ldots,Q\}$ and $m \neq n$ for a SIS with given state space $\mathcal{S} = \{\boldsymbol{s}_1, \boldsymbol{s}_2, \ldots, \boldsymbol{s}_Q\}$ encodes a failure transition, if*

1. $|\mathrm{Diff}(\boldsymbol{s}_m, \boldsymbol{s}_n)| = 1, i \in \mathrm{Diff}(\boldsymbol{s}_m, \boldsymbol{s}_n)$, *i.e., state space vectors $\boldsymbol{s}_m$ and $\boldsymbol{s}_n$ differ by only one index i which is the index of the failing component*
2. $\boldsymbol{s}_n[i] = OK$, *i.e., component i must be in OK state before the failure*
3. $(\boldsymbol{s}_m[i] \in \{DU, DUP, DN\}) \vee (\boldsymbol{s}_m[i] \in \{DDU, AU\} \wedge i \in \mathrm{Inh}(\boldsymbol{s}_m))$
$\vee (\boldsymbol{s}_m[i] \in \{DDR, AR\} \wedge i \notin \mathrm{Inh}(\boldsymbol{s}_m))$, *i.e., perform a transition to a failure state*

hold. Then the transition probability is

$$\boldsymbol{A}[m][n] = \begin{cases} \lambda_{du,i}\Delta t & \text{if } \boldsymbol{s}_m[i] = DU, \\ \lambda_{dup,i}\Delta t & \text{if } \boldsymbol{s}_m[i] = DUP, \\ \lambda_{dn,i}\Delta t & \text{if } \boldsymbol{s}_m[i] = DN, \\ \lambda_{dd,i}\Delta t & \text{if } (\boldsymbol{s}_m[i] = DDU \wedge i \in \mathrm{Inh}(\boldsymbol{s}_m)), \\ \lambda_{ddx,i}\Delta t & \text{if } (\boldsymbol{s}_m[i] = DDR \wedge i \in \mathrm{Inh}(\boldsymbol{s}_m)), \\ \lambda_{dd,i} + \lambda_{ddx,i}\Delta t & \text{if } (\boldsymbol{s}_m[i] = DDR \wedge i \notin \mathrm{Inh}(\boldsymbol{s}_m)), \\ \lambda_{a,i}\Delta t & \text{if } (\boldsymbol{s}_m[i] = AU \wedge i \in \mathrm{Inh}(\boldsymbol{s}_m)), \\ \lambda_{a,i}\Delta t & \text{if } (\boldsymbol{s}_m[i] = AR \wedge i \notin \mathrm{Inh}(\boldsymbol{s}_m)), \\ 0 & \text{else.} \end{cases}$$

A valid failure transition is thus given only, if a single component changes its state from OK to any failure state. This is ensured via the three preconditions given in def. 6.13.

The related failure states for *du*, *dup*, and *dn* failures are straight forward as they are not self-revealing. Further considerations are required for *dd*, *ddx*, and *a* failures.

The destination component state DDR state can be reached via two mechanisms. Firstly, *ddx* failures always lead to instant repair initiation, as the related external communication system is not prone to inhibition effects. Secondly, *dd* failures cause instant repair initiation, if they are not inhibited at the moment of occurrence. Therefore, the assigned failure rate is solely $\lambda_{ddx,i}$ if the destination state space vector $\boldsymbol{s}_m$ inhibits a possible *dd* failure, and $\lambda_{ddx,i} + \lambda_{dd,i}$ if not.

The destination component state DDU is relevant for inhibited dd failures only and thus reachable via $\lambda_{dd,i}$. Notice that an RF process with constant failure rate $\lambda_{dd,i}$ may lead to different component failure states, depending on the state of the rest of the SIS's components.

Active failures are treated similarly. In case of inhibition, a transition with component i changing its state from OK to AU is possible with failure rate $\lambda_{a,i}$. If not inhibited, then the destination state AR is reached immediately.

Failure transitions in the SMPMM encode the events $\boldsymbol{e_{F,du}}$, $\boldsymbol{e_{F,dup}}$, $\boldsymbol{e_{F,dn}}$, $\boldsymbol{e_{F,dd}}$, $\boldsymbol{e_{F,ddx}}$ and $\boldsymbol{e_{F,a}}$ from the CFSM from subsec. 5.2.3 (p. 73). The related CFSM states are contained in the considered state space vectors.

Definition 6.14 (Repair transition). *The transition matrix element $\boldsymbol{A}[m][n]$ with $\{m, n\} \in \{1, \ldots, Q\}$ and $m \neq n$ for a SIS with given state space $\mathcal{S} = \{\boldsymbol{s}_1, \boldsymbol{s}_2, \ldots, \boldsymbol{s}_Q\}$ encodes a repair transition, if*

1. $\exists! \, (i \in \mathrm{Diff}(\boldsymbol{s}_m, \boldsymbol{s}_n)) \, (\boldsymbol{s}_m[i] = OK \wedge \boldsymbol{s}_n[i] \in \{DDR, DR, AR\})$
2. $\exists \, (\mathcal{I}_{rel} = \mathrm{Inh}(\boldsymbol{s}_n) \setminus \mathrm{Inh}(\boldsymbol{s}_m)) \, ((\mathrm{Inh}(\boldsymbol{s}_m) \subseteq \mathrm{Inh}(\boldsymbol{s}_n)) \wedge (|\mathcal{I}_{rel}| \leq 1))$
   $\exists \, (\mathcal{I}_{rep} = \mathrm{Diff}(\boldsymbol{s}_m, \boldsymbol{s}_n) \setminus (\mathcal{I}_{rel} \cup i)) \, (\mathcal{I}_{rep} \subseteq \mathrm{MG}(c_i))$
3. $\forall \, (j \in \mathcal{I}_{rel}) \, ((\boldsymbol{s}_n[j] = DDU \wedge \boldsymbol{s}_m[j] = DDR) \vee (\boldsymbol{s}_n[j] = AU \wedge \boldsymbol{s}_m[j] = AR))$
4. $\forall \, (j \in \mathcal{I}_{rep}) \, ((\boldsymbol{s}_n[l] = DDU \wedge \boldsymbol{s}_m[l] = DDR) \vee (\boldsymbol{s}_n[l] = DU \wedge \boldsymbol{s}_m[l] = DR) \vee$
   $(\boldsymbol{s}_n[l] = DUP \wedge \boldsymbol{s}_m[l] = DR) \vee (\boldsymbol{s}_n[l] = AU \wedge \boldsymbol{s}_m[l] = AR))$
5. $\forall \, (j \in \mathcal{I}_{\overline{rep}}) \, (\boldsymbol{s}_n[l] \in \{DN, AR, DDR, DR\})$, *where*
   $\mathcal{I}_{\overline{rep}} = \mathrm{MG}(c_i) \setminus \mathrm{Diff}(\boldsymbol{s}_m, \boldsymbol{s}_n)$

hold, where $c_i \in \mathcal{COM}$ is the repaired component, and $\mathrm{MG}(\cdot)$ is the maintenance group function according to def. 5.24 (p. 98). Then the transition probability is

$$\boldsymbol{A}[m][n] = \begin{cases} \mu_{ddr,i}\Delta t & \text{if } \boldsymbol{s}_n[i] = DDR, \\ \mu_{dr,i}\Delta t & \text{if } \boldsymbol{s}_n[i] = DR, \\ \mu_{ar,i}\Delta t & \text{if } \boldsymbol{s}_n[i] = AR. \end{cases}$$

For repair processes, the same principles like for failure processes hold: only a single component may perform an FR process with a related repair rate per discrete time step. Additionally, the completion of an FR process may cause the generation of further events for other components, forcing them to change their internal state instantly. According to the CFSM from subsec. 5.2.3 (p. 73) the events $\boldsymbol{e_{R,ar}}$, $\boldsymbol{e_{R,dr}}$ and $\boldsymbol{e_{R,ddr}}$ are fired if a component returns back to its OK state. At the same time, this state transition may cause the event $\boldsymbol{e_{!inhibited}}$ to a component with inhibited failure that gets now revealed as the inhibiting component is functional again. Finally, $\boldsymbol{e_{mgr}}$ is generated for each member of the repaired component's maintenance group upon repair completion. Further failures in the group get instantly revealed. All of these events are encoded in the SMPMM's repair transitions.

A valid repair transition has origin and target state space vector differ by indices according to $\text{Diff}(\boldsymbol{s}_m, \boldsymbol{s}_n)$. Only one of the hereby represented component states is allowed to be the component that has its repair completed and thus performs a component state transition from *AR*, *DR* or *DDR* back to *OK* (see criterion 1). The related index is i.

$\mathcal{I}_{rel}$ is the set of release candidates (see criterion 2). In a valid repair transition, it describes all components that are inhibited in the origin state and not inhibited anymore in the destination state. According to property 6.1 (p. 113), $\mathcal{I}_{rel}$ contains one or no component. Notice that the transition is only valid according to def. 6.14, if no additional inhibitions occur due to the repair completion (which would be a natural contradiction to the idea of an FR process). This demand is satisfied by the requirement $\text{Inh}(\boldsymbol{s}_m) \subseteq \text{Inh}(\boldsymbol{s}_n)$. Criterion 3 demands that a released components needs to perform a transition from state *DDU* to *DDR* or from *AU* to *AR*. Other transitions do not make sense and thus lead to the invalidation of the criteria. The instant change of the released component's internal state depicts the implementation of the $\boldsymbol{e}_{!inhibited}$ event for the CFSM.

$\mathcal{I}_{rep}$ is the set of repair initiated components (reconsider criterion 2). In a valid repair transition, it describes all components that are not the repaired component (index i) and not the potentially released component, but components that change their internal state from either *DU*, *DUP*, *DDU* or *AU* to the respective repair states due to the maintenance group effect (see criterion 4). Other target component states obviously lead to the invalidation of the criteria. Hence, criterion 4 transition is the implementation of the $\boldsymbol{e}_{mgr}$ event.

$\mathcal{I}_{\overline{rep}}$ is the set of not repair initiated components. In a valid repair transition, it describes components in the maintenance group of the repaired component (index i) that do not change their internal state from origin to target state space vector. This is only tolerable, if said components are in a repair state already, i.e., *DR*, *DDR*, *AR*, or are in a non-repairable state that cannot be left due to the maintenance group effect: the *DN* state.

The diagonal elements in MM transition matrices are the recursive transitions, i.e., transitions leading back to their origin state. They can be calculated as 1 minus all probabilities for leaving the considered state, rendering these transitions easy to describe:

Definition 6.15 (Recursive transition). *The transition matrix element* $\boldsymbol{A}[m][n]$ *with* $\{m, n\} \in \{1, \ldots, Q\}$ *for a SIS with given state space* $\mathcal{S} = \{\boldsymbol{s}_1, \boldsymbol{s}_2, \ldots, \boldsymbol{s}_Q\}$ *encodes a recursive transition, if* $m = n$*. Then*

$$\boldsymbol{A}[m][n] = 1 - \sum_{l \neq n} \boldsymbol{A}[l][n].$$

Four short examples shall recapitulate the basic concept of transition matrix generation.

Example 6.5 (Transition matrix generation 1). Reconsider the ASMLG from fig. 5.11 (p. 97). The component state configuration vector specifies the order of the components in the state space vectors:

$$\overline{\boldsymbol{\sigma}}_0 := (\sigma_{\text{barrier01}}(t_0)\ \sigma_{\text{sensor}}(t_0)\ \sigma_{\text{barrier02}}(t_0))^\top .$$

It is assumed that no maintenance groups are specified. The first investigation considers a potential transition from $\boldsymbol{s}_n = (OK\ OK\ OK)^\top$ to $\boldsymbol{s}_m = (OK\ DU\ OK)^\top$.

- $\mathrm{Diff}(\boldsymbol{s}_m, \boldsymbol{s}_n) = \{2\}, |\mathrm{Diff}(\boldsymbol{s}_m, \boldsymbol{s}_n)| = 1$
- $\boldsymbol{s}_m[2] = DU$
- $\boldsymbol{s}_n[2] = OK$

Therefore, the described transition is possible and has to be classified as failure transition with probability $\lambda_{du,2}\Delta t$.

Example 6.6 (Transition matrix generation 2)**.** Consider a potential transition from $\boldsymbol{s}_n = (DU\ OK\ DU)^\top$ to $\boldsymbol{s}_m = (DU\ DDR\ DU)^\top$.

- $\mathrm{Diff}(\boldsymbol{s}_m, \boldsymbol{s}_n) = \{2\}, |\mathrm{Diff}(\boldsymbol{s}_m, \boldsymbol{s}_n)| = 1$
- $\mathrm{Inh}(\boldsymbol{s}_m) = \{2\}, \boldsymbol{s}_n(2) \notin \{DDU, AU\}$

The described transition is not possible, as a *dd* failure would be inhibited and thus require a transition of component 2 from OK to DDU instead of DDR.

Example 6.7 (Transition matrix generation 3)**.** Consider a potential transition from $\boldsymbol{s}_n = (DU\ DDU\ DR)^\top$ to $\boldsymbol{s}_m = (DU\ DDR\ OK)^\top$.

- $\mathrm{Diff}(\boldsymbol{s}_m, \boldsymbol{s}_n) = \{2, 3\}$
- $\{i \mid i \in \mathrm{Diff}(\boldsymbol{s}_m, \boldsymbol{s}_n) \wedge \boldsymbol{s}_m[i] = OK \wedge \boldsymbol{s}_n[i] \in \{DDR, DR, AR\}\} = \{3\}$
- $\mathcal{I}_{rel} = \mathrm{Inh}(\boldsymbol{s}_n) \setminus \mathrm{Inh}(\boldsymbol{s}_m) = \{2\} \setminus \{\} = \{2\}$ and $\{\} \subseteq \{2\}$ and $|\mathcal{I}_{rel}| = 1$
- $\mathcal{I}_{rep} = \mathrm{Diff}(\boldsymbol{s}_m, \boldsymbol{s}_n) \setminus (\mathcal{I}_{rel} \cup i) = \{2, 3\} \setminus (\{2\} \cup \{3\}) = \{2, 3\} \setminus \{2, 3\} = \{\}$ and $(\mathcal{I}_{rep} = \{\}) \subseteq (\mathrm{MG}(c_i) = \{\})$
- $\mathcal{I}_{\overline{rep}} = \mathrm{MG}(c_i) \setminus \mathrm{Diff}(\boldsymbol{s}_m, \boldsymbol{s}_n) = \{\} \setminus \{2, 3\} = \{\}$
- $\boldsymbol{s}_n[2] = DDU \wedge \boldsymbol{s}_m[2] = DDR$

Therefore, the described transition is possible and has to be classified as repair transition with probability $\mu_{dr,3}\Delta t$.

6.3.4 Phase transition matrices

In def. 4.1 (p. 56), a basic MPMM has been defined. Phase transitions have been implemented using a reset matrix $\boldsymbol{R}$. This matrix serves the purpose of providing information on how to 'shift' probability across the probability distribution vector at the time of a phase transition. For a given MPMM, the matrix $\boldsymbol{R}$ can be constructed easily, if all components are, e.g., proof tested simultaneously, resulting in a single reset matrix. As the ASML approach enables the safety engineer to specify arbitrary proof test times, replacement times etc. component individually, the number of required reset matrices increases significantly. If, e.g., one component is exclusively proof tested after one year, this requires a reset matrix $\boldsymbol{R}_1$. If after two years the same component is tested again, but together with a second component, then another reset matrix $\boldsymbol{R}_2 \neq \boldsymbol{R}_1$ is required. All of these matrices have to be generated from the available ASMLSIS. The subsequently provided generator functions make use of a remarkable property of reset matrices:

Property 6.2 (Reset matrix multiplication). *Be $\boldsymbol{R}_1$, $\boldsymbol{R}_2$, ..., $\boldsymbol{R}_n$ reset matrices describing the probability shifting procedure for the respective single components c_1, c_2, ..., c_n. Then $\boldsymbol{M} = \boldsymbol{R}_1\boldsymbol{R}_2 \cdots \boldsymbol{R}_n$ describes the probability shifting process, where all n components receive maintenance simultaneously.*

It is therefore possible to construct reset matrices for arbitrary simultaneously maintained components by multiplying the appropriate reset matrices for the individual components. The individual reset matrices are called 'component reset matrices' ($\boldsymbol{R}$) from now on. The resulting matrices describing all component maintenance procedures executed simultaneously at the time of a phase transition are called 'phase transition matrices' ($\boldsymbol{M}$) from now on.

Definition 6.16 (Component reset matrices). *For each component $c_i \in \mathcal{COM}$ three $Q \times Q$ reset matrix can be constructed. These encode how the probabilities in the SMPMM's probability distribution vector are shifted if a component receives maintenance due to a proof test or a group repair ($\boldsymbol{R}_{pt,i}$), a partial proof test ($\boldsymbol{R}_{ppt,i}$), or due to expired useful lifetime ($\boldsymbol{R}_{mt,i}$).*

$$\boldsymbol{R}_{pt,i}[m][n] = \begin{cases} 1 & \textit{if } m = n \land \boldsymbol{s}_m[i] \in \{OK, DR, DDR, AR, DN\}, \\ 1 & \textit{if } m \neq n \land \boldsymbol{s}_n[i] \in \{DU, DUP\} \\ & \land \boldsymbol{s}_m[i] = DR \land \forall (j \neq i) (\boldsymbol{s}_m[j] = \boldsymbol{s}_n[j]), \\ 1 & \textit{if } m \neq n \land \boldsymbol{s}_n[i] \in \{DDU\} \\ & \land \boldsymbol{s}_m[i] = DDR \land \forall (j \neq i) (\boldsymbol{s}_m[j] = \boldsymbol{s}_n[j]), \\ 1 & \textit{if } m \neq n \land \boldsymbol{s}_n[i] \in \{AU\} \\ & \land \boldsymbol{s}_m[i] = AR \land \forall (j \neq i) (\boldsymbol{s}_m[j] = \boldsymbol{s}_n[j]), \\ 0 & \textit{else.} \end{cases}$$

$$\boldsymbol{R}_{ppt,i}[m][n] = \begin{cases} 1 & \textit{if } m = n \land \boldsymbol{s}_m[i] \in \{OK, DR, DDR, AR, DN, DU, AU, DDU\}, \\ 1 & \textit{if } m \neq n \land \boldsymbol{s}_n[i] \in \{DUP\} \\ & \land \boldsymbol{s}_m[i] = DR \land \forall (j \neq i) (\boldsymbol{s}_m[j] = \boldsymbol{s}_n[j]), \\ 0 & \textit{else.} \end{cases}$$

$$
\boldsymbol{R}_{mt,i}[m][n] = \begin{cases} 1 & \textit{if } m = n \land \boldsymbol{s}_m[i] \in \{OK, DR, DDR, AR\}\,, \\ 1 & \textit{if } m \neq n \land \boldsymbol{s}_n[i] \in \{DU, DUP, DN\} \\ & \land \boldsymbol{s}_m[i] = DR \land \forall\,(j \neq i)\,(\boldsymbol{s}_m[j] = \boldsymbol{s}_n[j])\,, \\ 1 & \textit{if } m \neq n \land \boldsymbol{s}_n[i] \in \{DDU\} \\ & \land \boldsymbol{s}_m[i] = DDR \land \forall\,(j \neq i)\,(\boldsymbol{s}_m[j] = \boldsymbol{s}_n[j])\,, \\ 1 & \textit{if } m \neq n \land \boldsymbol{s}_n[i] \in \{AU\} \\ & \land \boldsymbol{s}_m[i] = AR \land \forall\,(j \neq i)\,(\boldsymbol{s}_m[j] = \boldsymbol{s}_n[j])\,, \\ 0 & \textit{else.} \end{cases}
$$

For all three matrices, $\{\boldsymbol{s}_m, \boldsymbol{s}_n\} \in \mathcal{S}$.

As introduced in subsec. 4.4.5 (p. 55), a reset matrix entry $\boldsymbol{R}_{(\cdot),i}[m][n] = 1$ denotes a probability shifting process within the MM from state n to state m. 1 entries on the diagonal refer to MM states which are not affected by the shifting process. For the proof test related reset matrix $\boldsymbol{R}_{pt,i}$ this holds for all entries where the considered component i is already in a repair state, i.e., $\{\boldsymbol{s}_m[i], \boldsymbol{s}_n[i]\} \subset \{DR, DDR, AR\}$, where it contains a non-detectable failure state DN, or where it is in OK state. In all other cases, the generator function ensures that shifts are performed between states with identical component states besides the considered component i and that said component i is transferred from an undetected failure state (DU, DDU, DUP, AU) to the respective matching repair state.

The other component reset matrices differ from $\boldsymbol{R}_{pt,i}$ by the types of undetected failures to be revealed during the test or replacement procedure. $\boldsymbol{R}_{mt,i}$ for instance allows a shifting process of components in DN state to the respective DR state.

The component reset matrices can be combined in order to retrieve phase transition matrices describing all probability shifting procedures required at a phase transition. The following definition utilizes the sets $\mathcal{PT}$, $\mathcal{PPT}$ and $\mathcal{MT}$ specified for each component in an ASML description. The continuous times provided in the sets are discretized in order to fit them into the discrete time SMPMM domain.

Definition 6.17 (Phase transition set). *The set of phase transitions $\mathcal{T}$ can be constructed from the component individual reset matrices $\boldsymbol{R}_{pt,i}$, $\boldsymbol{R}_{ppt,i}$ and $\boldsymbol{R}_{mt,i}$ as defined in def. 6.16, the component individual mission time set $\mathcal{MT}$ as well as the proof test interval set $\mathcal{PT}$ and the partial proof test set $\mathcal{PPT}$ (all three defined in def. 5.11 (p. 83)):*

$$
\mathcal{T} = \left\{ (k_t, \boldsymbol{M}_t) \;\middle|\; \begin{array}{l} k_t < k_{mt} \\ \land\, k_t \in \bigcup\limits_{i=1}^{N_{SIS}} (\mathcal{PT}'_i \cup \mathcal{PPT}'_i \cup \mathcal{MT}'_i) \\ \land\, \boldsymbol{M}_t = \prod\limits_{a|k_t \in \mathcal{PPT}'_a} \boldsymbol{R}_{pt,a} \prod\limits_{b|k_t \in \mathcal{PPT}'_b} \boldsymbol{R}_{ppt,b} \prod\limits_{c|k_t \in \mathcal{MT}'_c} \boldsymbol{R}_{mt,c} \end{array} \right\},
$$

with

$$\mathcal{PT}'_i = \{pt'_{1,i}, pt'_{2,i}, \ldots\} \ \textit{with}\ pt'_{j,i} = \text{floor}\left(\frac{pt_{j,i}}{\Delta t}\right) \ \textit{and}\ pt_{j,i} \in \mathcal{PT}_i,$$

$$\mathcal{PPT}'_i = \{ppt'_{1,i}, ppt'_{2,i}, \ldots\} \ \textit{with}\ ppt'_{j,i} = \text{floor}\left(\frac{ppt_{j,i}}{\Delta t}\right) \ \textit{and}\ ppt_{j,i} \in \mathcal{PPT}_i,$$

$$\mathcal{MT}'_i = \{mt'_{1,i}, mt'_{2,i}, \ldots\} \ \textit{with}\ mt'_{j,i} = \text{floor}\left(\frac{mt_{j,i}}{\Delta t}\right) \ \textit{and}\ mt_{j,i} \in \mathcal{MT}_i,$$

where floor $(\cdot)$ *is the 'round down' operator.*

Sets $\mathcal{PT}'_i$, $\mathcal{PPT}'_i$, and $\mathcal{MT}'_i$ describe the discrete time steps at which proof tests, partial proof tests or mission time replacements are conducted. They are constructed from the original component individual sets via division of each element by the step size Δt and rounding the result down to the respective nearest integer.

For each potential discrete time step k_t with deterministic SIS interaction, the sets $\mathcal{PT}'_i$, $\mathcal{PPT}'_i$, and $\mathcal{MT}'_i$ for all components i are investigated whether a component individual reset process is required at k_t. All related reset matrices are multiplied in order to retrieve the phase transition matrix $\boldsymbol{M}_t$ which represents *all* component shifting processes to be performed at k_t.

Example 6.8 (Phase transition generation)**.** A SIS consists of three components com01, com02, and com03 with

$$\begin{aligned}&\mathcal{PT}_{\text{com01}} = \{1\,\text{y}, 2\,\text{y}\}, \mathcal{PT}_{\text{com02}} = \{1\,\text{y}, 3\,\text{y}\}, \mathcal{PT}_{\text{com03}} = \{2\,\text{y}, 3\,\text{y}\},\\ &\mathcal{MT}_{\text{com01}} = \{3\,\text{y}\}, \mathcal{MT}_{\text{com02}} = \{2\,\text{y}\}, \mathcal{MT}_{\text{com03}} = \{1\,\text{y}\}.\end{aligned}$$

Then with $\Delta t = 1\,\text{h}$ the set of phase transitions is

$$\mathcal{T} = \left\{(k_1, \boldsymbol{M}_1)^\top, (k_2, \boldsymbol{M}_2)^\top, (k_3, \boldsymbol{M}_3)^\top\right\},$$

where

$$\begin{aligned}k_1 &= 8760, k_2 = 17520, k_3 = 26280,\\ \boldsymbol{M}_1 &= \boldsymbol{R}_{pt,\text{com01}} \boldsymbol{R}_{pt,\text{com02}} \boldsymbol{R}_{mt,\text{com03}},\\ \boldsymbol{M}_2 &= \boldsymbol{R}_{pt,\text{com01}} \boldsymbol{R}_{mt,\text{com02}} \boldsymbol{R}_{pt,\text{com03}},\\ \boldsymbol{M}_3 &= \boldsymbol{R}_{mt,\text{com01}} \boldsymbol{R}_{pt,\text{com02}} \boldsymbol{R}_{pt,\text{com03}}.\end{aligned}$$

Chapter 7

Optimization approaches

7.1 Overview

A mathematical framework with the intention of supporting the design process of safety instrumented systems needs to consider implementational constraints. The provided markovian approach suffers from the size of its state space. This might be a minor problem during the construction phase of the model (i.e., the transformation from an ASML description to the SMPMM), but a severe problem at runtime, while executing the recursive Chapman-Kolmogorov extension from def. 6.3 (p. 107). Here, memory as well as time consumption rise exponentially with the number of represented components.

Example 7.1 (Implementation problem for the transition matrix of a SIS with $N_{SIS} = 10$ components). According to the generator functions, the size of the state space equals $Q = |\mathcal{CS}|^{|N_{SIS}|}$ (def. 6.6 (p. 107)). Assuming a memory consumption of 8 bytes for a standard floating point variable with double precision, a simple SIS with 10 single components results in a transition matrix of around $|\mathcal{CS}|^{N_{SIS} \cdot 2} \cdot 8 \cdot 10^{-6} = 9.7 \cdot 10^{13}$ megabytes. A matrix multiplication has to be performed each discrete time step k. In a standard implementation that is Q^3 element multiplications per regular matrix multiplication (disregarding additions) and thus $2Q^3$ for each phase transition [Wag03]. On a quad core machine with 3,2 GHz CPU ($\approx$ 30 GFLOPS) and one FLOP per multiplication, this results in $|\mathcal{CS}|^{N_{SIS} \cdot 3} \cdot \frac{1}{30 \cdot 10^9} = 9^{30} \cdot \frac{1}{30 \cdot 10^9}\,\text{s} \approx 5.5 \cdot 10^{10}$ years per standard matrix multiplication. The calculation of a 30 year mission time does not even make sense.

The plain generator functions deliver calculation models that can be executed on smaller CPUs in a reasonable computation time for up to around 6 components. Larger systems can easily be executed at electronic data processing centers (EDPCs). But as extended modifications to the algorithms are required in order to support the hyper-threading, i.e., parallel computing abilities, a more detailed look on the models themself is worthwhile. With rather simple methods, the computational effort required for executing and storing the calculation models can be decreased significantly.

The subsequently provided approaches deal with the optimization challenge from different starting points. The ASMLG decomposition, the state space reduction, and the sparse matrix implementation are lossless, i.e., do not have an impact on the calculation result. They are therefore preferred methods and described in detail in secs. 7.2 to 7.4. The aggregation of channel components as well as state space compression methods are always at the expense of accuracy. They are briefly introduced in sec. 7.5.

Notice that all presented approaches need to be applied generically in order to fit into the overall ASML concept. The actually implemented features are outlined in detail by providing the related generator functions or at least parts of them. This should give a brief introduction into how to integrate the various approaches into the workflow.

7.2 ASMLG decomposition

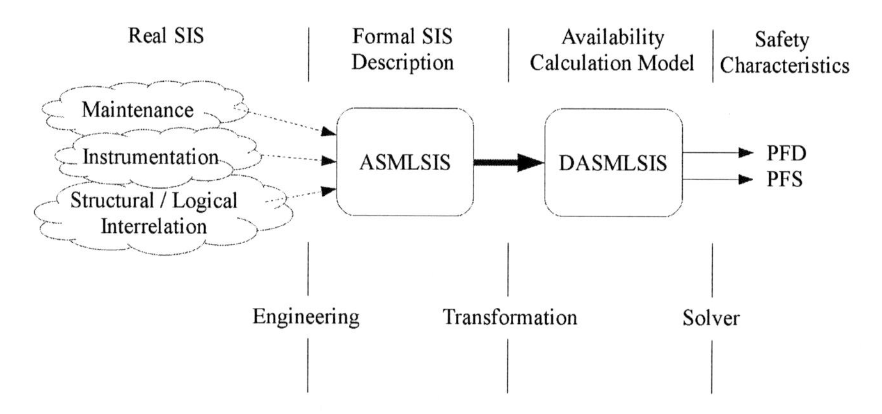

Figure 7.1: Flow diagram of the ASML decomposition approach

The combinatorial approaches in sec. 4.2 (p. 42) and sec. 4.3 (p. 45) showed, how independent (and thus stochastically independent) unavailability functions can be combined via suitable combinatorial terms in order to retrieve the SIS's unavailability function. These approaches were sorted out when a choice for the basic modeling approach for the considered class of SIS was made (sec. 4.6 (p. 66)). The main reason was the missing capability of reproducing interdependencies among components.

As far as at least parts of the SIS described in ASML are independent from each other, a hierarchical decomposition of an ASMLG is possible. The basic concept is to split an ASMLG into several subsystems that - together with subsystem interdependency information - form a decomposed ASML SIS (DASMLSIS, see fig. 7.1). The most important restriction for the decomposition is - as said above - that all subsystems have to be stochastically independent from each other. In other words: the respective subsystem unavailability must not depend on the unavailabilities of other subsystems via, e.g., maintenance group triggered repair initiation or inhibition effects. For each of the subsystems, the unavailability measures over the discrete time are calculated according to def. 6.4

(p. 107). The resulting unavailability curves are recombined using suitable combinatorial functions to be provided in this subsection. The results are the unavailabilities of the original SIS. All required definitions are similar to the ones given throughout the chapter on ASML graphs (chapter 5).

The mentioned procedure resembles the combinatorial approach as introduced in subsec. 4.3.3 (p. 47), where individual component unavailability curves have been combined in order to retrieve the system's unavailability.

The benefit from an ASMLSIS decomposition is the dramatic reduction of the size of the state space. Instead of calculating one large markov model, the required unavailabilities are retrieved by calculating several small markov models, one for each subsystem. It is important to point out that the safety engineer providing an ASML description of a SIS needs not to be aware of independencies among the system. The optimization as provided in this section automatically detects independencies and performs the decomposition. Therefore, no additional expert knowledge is required.

A closing example on page 132 will outline the benefit in numbers.

7.2.1 Decomposed ASML graph

The first step is to introduce a new type of graph, the decomposed ASML graph.

Figure 7.2: Vertex types in decomposed ASML graphs

Definition 7.1 (Decomposed ASML graph). *The tuple*

$$\mathcal{DASMLG} = \{\mathcal{DV}, \mathcal{DE}\}$$

is called a decomposed ASML (di)graph (DASMLG). $\mathcal{DV}$ is a set of vertices. $\mathcal{DE}$ is a set of edges according to

$$\mathcal{DE} \subset (\mathcal{DV} \times \mathcal{DV}).$$

The interpretation of an edge $(dv_1, dv_2) \in \mathcal{DE}$ is as follows: dv_2's safety related/operational availability depends on dv_1's safety related/operational availability. Thus, the desired safety characteristics of dv_2 depend on all vertices $\{dv \mid (dv, dv_2) \in \mathcal{DE}\}$. A DASMLG is part of the optimized description of a SIS (see def. 7.8 (p. 127).

The attributized vertices $\mathcal{DV}$ of an DASMLG for a given SIS are of two mutually different types:

$$\mathcal{DV} = \mathcal{SUB} \cup \mathcal{CSEQ}, \textit{with } \mathcal{SUB} \cap \mathcal{CSEQ} = \emptyset.$$

$\mathcal{SUB}$ is a set of subsystems (see def. 7.3), $\mathcal{CSEQ}$ is a set of combinatorial sequences (see def. 7.4). Both vertex types are represented by specific symbols as shown in fig. 7.2.

Some remarks on def. 7.1:

- The additional parameterization as introduced in the related definitions may be denoted in the DASMLG or left out for simplification purposes. However, the related mathematical objects must always be fully parameterized.
- Both the set of vertices and the set of edges, are obviously restricted, since, e.g., $\mathcal{DE} \nsubseteq (\mathcal{DV} \times \mathcal{DV})$ (compare def. 7.2).
- Graph traversing is defined according to the respective definitions for ASMLGs in subsec. 5.3.1 (p. 81).

Definition 7.2 (DASMLG restrictions). *The set of edges $\mathcal{DE} \subset (\mathcal{DV} \times \mathcal{DV})$ consisting of tuples (dv_1, dv_2) with $\{dv_1, dv_2\} \subseteq \mathcal{DV}$ for an arbitrary DASMLG is restricted according to*

1. $\mathcal{DE} \subseteq ((\mathcal{DV} \times \mathcal{DV}) \setminus (\mathcal{DV} \times \mathcal{SUB}))$, *i.e., subsystems have only abducent edges.*
2. $(\exists! dv_{SIS} \in \mathcal{DV})\,((\nexists\,(dv_1, dv_2) \in \mathcal{DE})\,(dv_1 = dv_{SIS}))$, *i.e., in DASMLGs only a single vertex dv_{SIS} without abducent edge exists: the SIS's unavailability output.*

The layout of an arbitrary DASMLG is restricted according to

3. $\{p \mid \exists m \exists n\,((\bullet p[m] = p[n]\bullet) \wedge (m \leq n))\} \nsubseteq \mathcal{DP}$,

where $\mathcal{DP}$ is the set of all possible paths inside the DASMLG. This last restriction ensures loop-freeness.

With defs. 7.1 and 7.2, a first exemplary DASMLG may be investigated. Reconsider the SIS outlined in fig. 7.3 (p. 125). A related ASMLG is depicted in fig. 5.4 (p. 79). Finally, fig. 7.3 contains the DASMLG for the specified SIS. The original ASMLG has been decomposed into three subsystems and a single combinatorial sequence which also serves as the system's output. Each subsystem represents a part of the original ASMLG. The mathematical structure of a subsystem is as follows:

Definition 7.3 (Subsystem). *An arbitrary subsystem sub_i in the set of subsystems $\mathcal{SUB}$ of an DASMLG is defined as a one tuple according to*

$$sub_i \in \mathcal{SUB} = \mathcal{SUBSYSTEM}\,(SMPMM_i)\,,$$

where SMPMM is a safety multiphase markov model according to subsec. 6.1 (p. 106). Typically, the SMPMM related to a subsystem represents the calculation model for a stochastically independent part of the original ASMLG.

The three subsystems sub01, sub02, and sub03 in fig. 7.3 therefore represent the outlined parts of the original ASMLG. The subsequentially presented generator functions for DASMLGs (def. 7.11 (p. 131)) require to identify suitable parts of the ASMLG that may be used as subsystems. Afterwards, the SMPMMs for the identified parts of the ASMLG are to be generated and finally transferred into a subsystem vertex. The interaction among subsystem unavailabilities can be accomplished via combinatorial sequences:

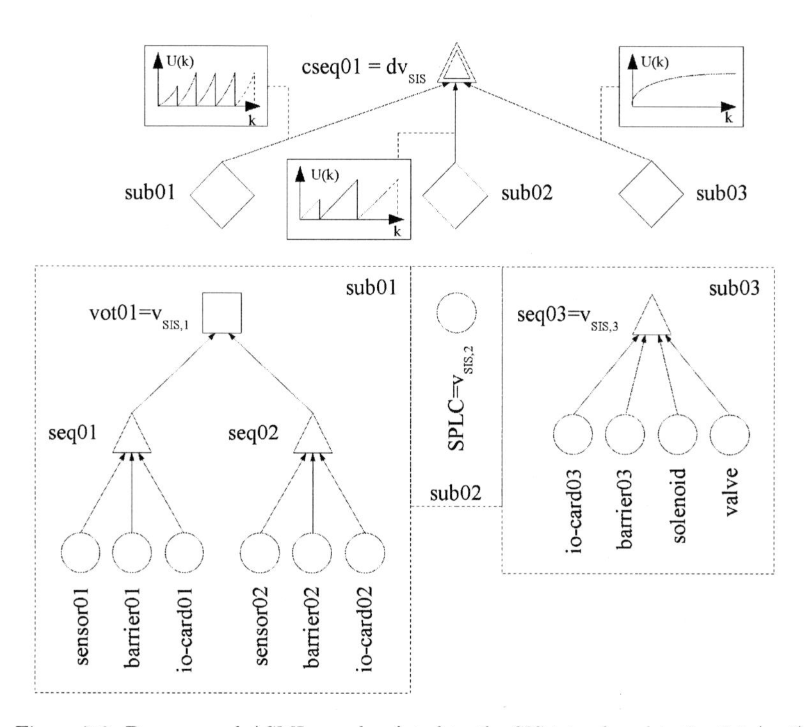

Figure 7.3: Decomposed ASML graph related to the SIS introduced in fig. 2.1 (p. 9)

Definition 7.4 (Combinatorial sequence). *An arbitrary combinatorial sequence cs_i in the set of combinatorial sequences $\mathcal{CSEQ}$ is a one-tuple*

$$cs_i \in \mathcal{CSEQ} = \mathcal{COMBINATORIAL-SEQUENCE}\left(\boldsymbol{iv}_i\right),$$

specifying the temporal order of vertices contributing to the forwarding process of a trip signal with their respective availabilities. $\boldsymbol{iv}_i$ with $\boldsymbol{iv}_i[j] \in \mathcal{DV}$ is a vector of input vertices, specifying the order of all N involved vertices. The length of this vector corresponds to the number of pre edges for the considered combinatorial sequence:

$$|\boldsymbol{iv}_i| = |\bullet cs_i| = N.$$

The pre vertices of the edges $\bullet cs_i$ are the elements contained in the input vertex vector:

$$\{dv1 \mid (dv1, dv2) \in \bullet cs_i\} = \{\boldsymbol{iv}_i[1], \boldsymbol{iv}_i[2], \ldots, \boldsymbol{iv}_i[N]\}.$$

$\boldsymbol{iv}[1]$ is the temporarily first, and $\boldsymbol{iv}[N]$ the temporarily last required vertex. The order of input vertices corresponds to the order of the considered vertices in DASMLGs from left to right (if not explicitly defined otherwise).

The definition for combinatorial sequences is very close to the one for ASML sequences in def. 5.13 (p. 85). The behavior is exactly the same. Differences can be found in the evaluation functions.

7.2.2 DASMLG evaluation

The evaluation of DASMLGs differs from standard ASMLGs, as the graph now consists of subsystems (represented by SMPMMs) and combinatorial sequences only. From the multiphase markov models, only discrete time dependent information can be retrieved. Therefore, the evaluation functions are also functions of time. Similar to sec. 5.3.6 (p. 89), a generalized evaluation function is provided first. It serves as an interface to the various vertex types utilized in an DASMLG.

Definition 7.5 (Generalized DASMLG evaluation)**.** *The set of generalized discrete time unavailability evaluation functions for arbitrary vertices $dv \in \mathcal{DV}$ is*

$$\begin{aligned}
&\mathrm{PFD} : (\mathcal{DV} \times \mathbb{N}_+) \mapsto \mathbb{R} \\
&:= \mathrm{PFD}(dv,k) = \begin{cases} \mathrm{PFD}_{\mathrm{SUB}}(dv,k) & \text{if } dv \in \mathcal{SUB}, \\ \mathrm{PFD}_{\mathrm{CSEQ}}(dv,k) & \text{if } dv \in \mathcal{CSEQ}. \end{cases} \\
&\mathrm{PFS} : (\mathcal{DV} \times \mathbb{N}_+) \mapsto \mathbb{R} \\
&:= \mathrm{PFS}(dv,k) = \begin{cases} \mathrm{PFS}_{\mathrm{SUB}}(dv,k) & \text{if } dv \in \mathcal{SUB}, \\ \mathrm{PFS}_{\mathrm{CSEQ}}(dv,k) & \text{if } dv \in \mathcal{CSEQ}. \end{cases}
\end{aligned}$$

$\mathrm{PFD}(dv,k)$ *denotes the probability of having vertex dv s-unavailable at time step k.* $\mathrm{PFS}(dv,k)$ *is the probability for the vertex to be o-unavailable and thus in a spurious trip state, where the predefined safety function has successfully, but unintendly been triggered.*

Definition 7.6 (Subsystem evaluation)**.** *The discrete time unavailability of a subsystem vertex sub_i in the set of subsystems $\mathcal{SUB}$ of an DASMLG can be evaluated according to*

$$\begin{aligned}
\mathrm{PFD}_{SUB}(sub_i,k) &= \mathrm{PFD}_{SMPMM_i}(k), \\
\mathrm{PFS}_{SUB}(sub_i,k) &= \mathrm{PFS}_{SMPMM_i}(k),
\end{aligned}$$

with all functions mapping $(\mathcal{SUB} \times \mathbb{N}_+) \mapsto \mathbb{R}$. $\mathrm{PFD}_{SMPMM_i}(k)$ and $\mathrm{PFS}_{SMPMM_i}(k)$ are the associated unavailabilities over time as defined in def. 6.4 (p. 107).

Definition 7.7 (Combinatorial sequence evaluation)**.** *The discrete time unavailability evaluation functions for an arbitrary combinatorial sequence cs_i in the set of combinatorial sequences $\mathcal{CSEQ}$ are defined as*

$$\begin{aligned}
\mathrm{PFD}_{CSEQ}(cs_i,k) &= \sum_{m=1}^{N} \left(\mathrm{PFD}(\boldsymbol{iv}_i[m],k)\,\mathrm{POK}(cs_i,m,k)\right), \\
\mathrm{PFS}_{CSEQ}(cs_i,k) &= \sum_{m=1}^{N} \left(\mathrm{PFS}(\boldsymbol{iv}_i[m],k)\,\mathrm{POK}(cs_i,m,k)\right).
\end{aligned}$$

All functions map $(\mathcal{CSEQ} \times \mathbb{N}_+) \mapsto \mathbb{R}$. $\mathrm{POK}(cs_i, m, k)$ *is defined according to*

$$\mathrm{POK}(cs_i, m, k) = \begin{cases} \prod_{n=m+1}^{N} (1 - \mathrm{PFS}(\boldsymbol{iv}_i[n], k) - \mathrm{PFD}(\boldsymbol{iv}_i[n], k)) & \textit{if } m < N, \\ 1 & \textit{else.} \end{cases}$$

The function $\mathrm{POK}(cs_i, m, k)$ ('probability of OK') denotes the probability for all input vertices with higher index than m to be neither s- nor o-unavailable. The product $\mathrm{PFD}(\boldsymbol{iv}_i[m], k)\,\mathrm{POK}(cs_i, m, k)$ for given m thus denotes the contribution of input vertex m to the safety related unavailability of the considered combinatorial sequence. The sum over all possible indices $1 \leq m \leq N$ delivers the overall combinatorial sequence unavailability.

The possible result of a DASMLG evaluation is outlined in fig. 7.3 (p. 125). The small boxes denote the subsystems' s-unavailabilities as functions over time. The DASMLG output vertex cseq01 $= dv_{SIS}$ aggregates all curves using the combinatorial formulas from def. 7.7 to retrieve the system's unavailability. Notice that this principle is very close to the combinatorial approaches including RBDs and FTs (see secs. 4.2 and 4.3). The derivation of the averaged measures is denoted in the following subsection.

7.2.3 DASML SIS

The DASMLG is an important part of the overall system, the DASMLSIS:

Definition 7.8 (DASMLSIS)**.** *A DASMLSIS is the basic mathematical structure, holding all required information about the safety instrumented function to be described. It is a three-tuple according to*

$$\mathcal{DASMLSIS}\left(k_{mt}, \underbrace{\mathcal{DV}, \mathcal{DE}}_{DASMLG}\right),$$

where k_{mt} *is the number of mission time steps,* $\mathcal{DV}$ *are the attributized vertices, and* $\mathcal{DE} \subset (\mathcal{DV} \times \mathcal{DV})$ *the edges of the related DASMLG (see def. 7.1 (p. 123)).*

With the DASMLSIS definition and the evaluation functions def. 7.7 and def. 7.6, the desired unavailability characteristics for the DASMLSIS under investigation can easily be retrieved:

Definition 7.9 (DASMLSIS unavailability)**.** *The discrete time unavailabilities of a DASML-SIS can be calculated according to*

$$\mathrm{PFD}(k) = \mathrm{PFD}(dv_{SIS}, k),$$
$$\mathrm{PFS}(k) = \mathrm{PFS}(dv_{SIS}, k).$$

Based on the unavailabilities over time, the averaged measures can be derived as

$$\text{PFD}_{avg} = \frac{1}{k_{mt}} \sum_{k=0}^{k_{mt}-1} \text{PFD}(k),$$

$$\text{PFS}_{avg} = \frac{1}{k_{mt}} \sum_{k=0}^{k_{mt}-1} \text{PFS}(k),$$

where k_{mt} is the number of discrete mission time steps of the related DASMLSIS.

7.2.4 Transformation: ASMLSIS to DASMLSIS

The first step towards a transformation from a given ASMLSIS to a related DASMLSIS is the identification of potential subsystems, i.e., stochastically independent groups of ASML vertices. The optimization approach presented in this section provides suitable transformation formulas for splitting ASMLGs at sequence vertices $s_i \in \mathcal{SEQ}$ as can be seen from fig. 7.3 (p. 125) in comparison with the original ASMLG in fig. 5.4 (p. 79). Splitting at components or voters is excluded from further considerations. The reason for this restriction derives from difficulties in formulating appropriate evaluation functions for other vertex types than sequences. Voters and components contain internal behavior of too high complexity as can reasonably be condensed into a single function (this complexity originally was the reason to choose a multiphase markov model as calculation approach). However, as almost all SISs consist of the system parts sensors, logic solver and final elements, the system output v_{SIS} is usually a sequence (compare all provided ASMLG examples). This renders the presented approach applicable for the largest group of systems.

The decomposition function is a supporting function. For a given sequence vertex $s_i \in \mathcal{SEQ}$, $\text{Dec}(s_i)$ evaluates whether the related inputs vertices may serve as output vertices of respective potential subsystems.

Definition 7.10 (Decomposition function). *The decomposition function* Dec *decides on whether a considered sequence $s_i \in \mathcal{SEQ}$ contains stochastically independent input vertices.*

$$\text{Dec} : \mathcal{SEQ} \mapsto \{true, false\} := \text{Dec}(s_i)$$
$$= \begin{cases} false & \text{if } \exists k_1 \exists k_2 \, (k_1 \neq k_2 \wedge (\text{PATH}(k_1, k_2) \neq \emptyset \vee \text{MAINT}(k_1, k_2) \neq \emptyset)), \\ true & \text{else,} \end{cases}$$

where

$$\text{PATH}(k_1, k_2) = \bigcup_{v_m \in \mathcal{V} | \text{EP}(v_m, \boldsymbol{iv}_i[k_1]) \neq \emptyset} v_m \cap \bigcup_{v_n \in \mathcal{V} | \text{EP}(v_n, \boldsymbol{iv}_i[k_2]) \neq \emptyset} v_n,$$

$$\text{MAINT}(k_1, k_2) = \bigcup_{c_m \in \mathcal{COM} | \text{EP}(c_m, \boldsymbol{iv}_i[k_1]) \neq \emptyset} \text{MG}(c_m) \cap \bigcup_{c_n \in \mathcal{COM} | \text{EP}(c_n, \boldsymbol{iv}_i[k_2]) \neq \emptyset} \text{MG}(c_n).$$

EP($\cdot$) *is the edge path function according to def. 5.8 (p. 82) and* MG($\cdot$) *is the maintenance group function according to def. 5.24 (p. 98).*

Notice that although the considered sequence s_i is not explicitly used in the function, its related input vector $\boldsymbol{iv}_i$ is.

The decomposition function verifies two properties sufficient for detecting stochastical independency among the input vertices of a given sequence. For arbitrary two different input vertices $\boldsymbol{iv}_i[k_1]$ and $\boldsymbol{iv}_i[k_2]$,

1. the respective unions of sets of vertices with existing path to the considered input vertex are constructed (PATH function). If the intersection of both vertex sets is not empty, then at least one vertex is connected to at least two input vertices of the considered sequence s_i.
2. the respective unions of sets of maintenance groups for components with valid paths to the considered input vertex are constructed (MAINT function). If the intersection of both index sets is not empty, then at least one component's maintenance group contains the index of a component connected to another input vertex of the investigated sequence.

In both cases the input vertices are not stochastically independent from each other, the decomposition is not possible. It is sufficient if only one of the two criteria is met. Two short examples shall demonstrate this principle.

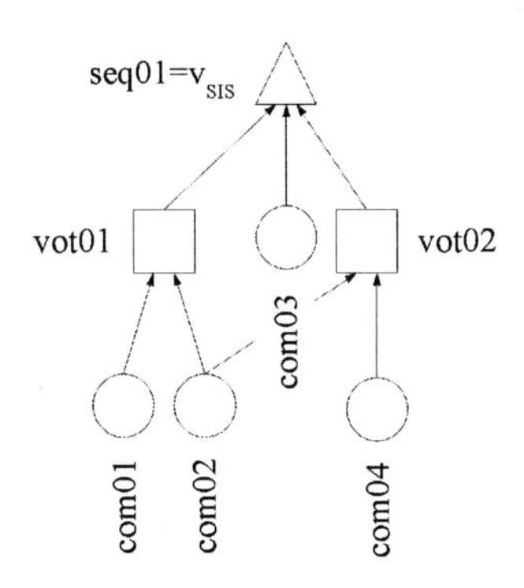

Figure 7.4: Non-decomposable ASMLG

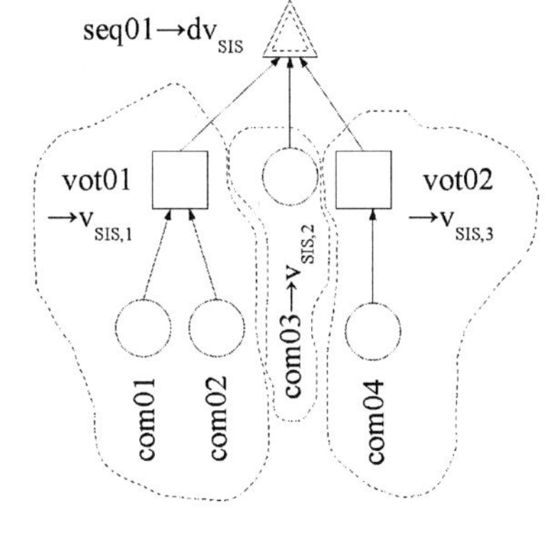

Figure 7.5: Decomposable ASMLG

Example 7.2 (ASMLG decomposition 1)**.** Consider fig. 7.4. It is to be investigated, whether the ASMLG can be decomposed at seq01. Therefore, Dec(seq01) is calculated. As the ASMLSIS related to the depicted ASMLG does not specify any maintenance groups, i.e., $\mathcal{MGS} = \emptyset$, the second criterion needs not to be regarded. The sets

$\mathrm{PATH}(k_1, k_2)$ can be constructed for all three possible input vector combinations from $\boldsymbol{iv}_{\text{seq01}} = (\text{vot01 com03 vot02})^\top$ according to:

$$\begin{aligned}\mathrm{PATH}(1,2) &= \{\text{com01}, \text{com02}\} \cap \{\} = \{\} \\ \mathrm{PATH}(1,3) &= \{\text{com01}, \text{com02}\} \cap \{\text{com02}, \text{com04}\} = \{\text{com02}\} \\ \mathrm{PATH}(2,3) &= \{\} \cap \{\text{com02}, \text{com04}\} = \{\}\end{aligned}$$

As $|\mathrm{PATH}(1,3)| \neq \emptyset$, the ASMLG cannot be split at seq01.

Example 7.3 (ASMLG decomposition 2). Consider fig. 7.5. It is to be investigated, whether the ASMLG can be decomposed at seq01. Therefore, Dec(seq01) is calculated. The ASMLSIS related to the depicted ASMLG specifies

$$\mathcal{MGS} = \{\{\text{com01}, \text{com02}\}\}$$

as maintenance group. The sets $\mathrm{PATH}(k_1, k_2)$ can be constructed for all three possible input vector combinations from $\boldsymbol{iv}_{\text{seq01}} = (\text{vot01 com03 vot02})^\top$ according to:

$$\begin{aligned}\mathrm{PATH}(1,2) &= \{\text{com01}, \text{com02}\} \cap \{\} = \{\} \\ \mathrm{PATH}(1,3) &= \{\text{com01}, \text{com02}\} \cap \{\text{com04}\} = \{\} \\ \mathrm{PATH}(2,3) &= \{\} \cap \{\text{com04}\} = \{\}\end{aligned}$$

Since all PATH sets are empty, the related criterion allows a decomposition at seq01.

$$\begin{aligned}\mathrm{MAINT}(1,2) &= \{\text{com01}, \text{com02}\} \cap \{\} = \{\} \\ \mathrm{MAINT}(1,3) &= \{\text{com01}, \text{com02}\} \cap \{\} = \{\} \\ \mathrm{MAINT}(2,3) &= \{\} \cap \{\} = \{\}\end{aligned}$$

The second criterion, using MAINT, also corroborates a decomposition of the ASMLG at seq01. Hence, a split can be performed such that com01, com02 and vot01 form a subsystem with vot01 as output vertex $v_{SIS,1}$ as well as com04 and vot02, forming a subsystem with vot02 as $v_{SIS,3}$. Component com03 alone is considered as a subsystem with com03 itself as output vertex. The subsystems are outlined with dotted lines in fig. 7.5.

Using the decomposition function, the generation of a DASMLSIS can be defined formally. Prior to the related definition, the course of actions shall be explained:

Starting at the considered ASMLGs output vertex v_{SIS}, the decomposability is checked by verifying that the considered vertex is a sequence and by evaluating the $\mathrm{Dec}(v_{SIS})$ function. If the ASMLG is decomposable at v_{SIS}, the N input vertices of v_{SIS} become the output vertices $v_{SIS,1}, \ldots, v_{SIS,N}$ of N new ASMLGs. These new ASMLGs already represent stochastically independent subgraphs. For each new system output $v_{SIS,j}$ it can be evaluated whether a further decomposition is possible. If not, each subgraph is transformed into a DASMLSIS by (among others) generating the respective SMPMMs and assigning these to subsystem vertices. A combinatorial sequence is generated for each ASML-sequence used for splitting the original ASMLG. subsystems and combinatorial sequences are linked together such that the original SIS behavior is kept and thus correctly reproduced.

Definition 7.11 (DASMLSIS generation). *Be* $s_i \in \mathcal{SEQ}_{orig}$ *with* $|\boldsymbol{iv}_i| = N$ *the output vertex* v_{SIS} *in the* $ASMLG_{orig}$ *of the description* $ASMLSIS_{orig}$ *of a SIS. A decomposition is possible, if* $\mathrm{Dec}(s_i) = true$. *The N subsystems can then be generated by first constructing N new ASMLSISs according to:*

$$ASMLSIS_j = (\tau_j, \mathcal{MGS}_j, \mathcal{V}_j, \mathcal{E}_j),$$

where $\tau_j = \tau_{orig}$, *i.e., the mission time of the original ASMLSIS. The set of maintenance groups for each* $ASMLSIS_j$ *can be generated via*

$$\mathcal{MGS}_j = \{\mathcal{MG} \mid \mathcal{MG} = \mathrm{MG}(c_{orig}) \wedge \mathrm{EP}(c_{orig}, \boldsymbol{iv}_i[j]) \neq \emptyset\}.$$

The respective sets of vertices and edges can be obtained via

$$\mathcal{V}_j = \{v_{orig} \in \mathcal{V}_{orig} \mid \mathrm{EP}(v_{orig}, \boldsymbol{iv}_i[j]) \neq \emptyset\}.$$
$$\mathcal{E}_j = \{e_{orig} \in \mathcal{E}_{orig} \mid e_{orig} \in \mathrm{EPD}(\boldsymbol{iv}_i[j])\}.$$

$\mathrm{EP}(\cdot)$ *is the edge path function (see def. 5.8 (p. 82)),* $\mathrm{EPD}(\cdot)$ *is the edge path destination function (see def. 5.9 (p. 82)). The two construction directives for* $\mathcal{V}_j$ *and* $\mathcal{E}_j$ *ensure that each* $ASMLG_j$ *has its single output vertex which is identical to an input vertex of the originally considered sequence* s_i*:*

$$v_{SIS,j} = \boldsymbol{iv}_i[j].$$

For each $ASMLSIS_j$ *a related* $SMPMM_j$ *can be generated by using the transformation formulas from sec. 6.3 (p. 107).*

The DASMLSIS can be constructed according to def. 7.8 (p. 127). The related DASML subsystems and combinatorial sequences, as well as a set of edges can be parameterized and added to the DASMLSIS according to

$$\mathcal{SUB} = \mathcal{SUB} \cup \{sub_1 = (SMPMM_1), sub_2 = (SMPMM_2), \ldots, sub_N = (SMPMM_N)\},$$
$$\mathcal{CSEQ} = \mathcal{CSEQ} \cup \{cs = (\boldsymbol{iv})\}, \text{where } \boldsymbol{iv} = (sub_1\ sub_2\ \ldots\ sub_N)^\top,$$
$$\mathcal{DE} = \mathcal{DE} \cup \{e_1 = (sub_1, cs), e_2 = (sub_2, cs), \cdots, e_N = (sub_N, cs)\}.$$

Instead of being transformed into a subsystem, each of s_i*'s input vertices* $\boldsymbol{iv}_i[j]$ *can be decomposed further if* $(\boldsymbol{iv}_i[j] = s_j) \in \mathcal{SEQ}_{orig}$ *and* $\mathrm{Dec}(s_j) = true$. *Then a combinatorial sequence* $cs_j \in \mathcal{CSEQ}$ *can be constructed from* s_j *and be applied to the DASMLG via an edge* $e = (cs_j, cs_i)$.

Finally, the DASMLSIS can be parameterized according to

$$DASMLSIS = \left(\mathrm{round}\left(\frac{\tau_{orig}}{\Delta t}\right), \mathcal{DV}, \mathcal{DE}\right).$$

This definition shall be accompanied by the related generator algorithm for the DASMLG. It makes use of recursion and is to be initially called with the original ASMLG and empty sets $\mathcal{DV}$, $\mathcal{DE}$.

If the first instance of the algorithm identifies the output vertex as 'not a sequence vertex' or not decomposable (line 1), then the ASMLG can be transformed into a DASMLG only in a whole. Notice, that in that case no edges are generated, as the algorithm ends in line 4. Edges are not necessary, as the DASMLG consists of only a single vertex, i.e., the subsystem represents the complete original ASMLG.

If the current ASMLG is decomposable, ASMLG_1 to ASMLG_N representing the stochastically independent parts of the original ASMLG are generated (line 6). For each ASMLG_j, the algorithm is recursively called (line 7). The respective return value is the DASMLG vertex representing ASMLG_j (via subsystem or combinatorial sequence).

A DASMLG combinatorial sequence cs_d is generated (line 11), representing the currently investigated ASML sequence s_i.

For each of the newly generated DASMLG vertices v an DASMLG edge $e = (v, cs_d)$ is generated, interconnecting v with the currently investigated ASML sequence s_i (line 12 - 14).

Input: ASMLG, sets $\mathcal{DV}$ and $\mathcal{DE}$
1 **if** $v_{SIS} \notin \mathcal{SEQ}$ **or** $\text{Dec}(v_{SIS}) = false$ **then**
2 | generate DASMLG subsystem *sub* from ASMLG;
3 | $sub_i \rightarrow \mathcal{DV}$;
4 | **return** *sub*;
5 **end**
6 decompose ASMLG into ASMLG_1 to ASMLG_N with $N = |\boldsymbol{iv}_i|$;
7 **foreach** $ASMLG_j$ **do**
8 | fetch DASMLG vertex $v = \text{DASMLGGeneratorAlgorithm}\,(\text{ASMLG}_j, \mathcal{DV}, \mathcal{DE})$;
9 | store DASMLG vertex for later use $\boldsymbol{iv}_d[j] = v$;
10 **end**
11 generate DASMLG combinatorial sequence $cs_d = (\boldsymbol{iv}_d) \rightarrow \mathcal{DV}$;
12 **for** $j = 1$ **to** $N = |\boldsymbol{iv}_i|$ **do**
13 | generate DASMLG edge $e = (\boldsymbol{iv}_d[j], cs_d) \rightarrow \mathcal{DE}$;
14 **end**
15 **return** cs_d

Algorithm 7.1: DASMLG Generator Algorithm

Example 7.4 (Demonstration of the optimization effect)**.** Reconsider ex. 7.1 (p. 121). If the 10 component SIS is supposed to consist of sensor part, logic solver, and final element part, then the output vertex v_{SIS} of the related ASMLG will be a sequence. It is likely that the three system parts are not stochastically dependent from each other. Hence, at least one decomposition step is possible. Potential further decompositions are disregarded for this example.

Then the components might be distributed to three subsystems with an amount of, e.g., 4 - 2 - 4 for sensors - logic solver - final elements. This leads to transition matrices of around 344 - 0.05 - 344, i.e., 688.1 megabytes in total. The calculation time for a single matrix multiplication decreases to 9.4 - $1.7 \cdot 10^{-5}$ - 9.4 s, i.e., 18.8 s in total. With that, the calculation time for a 30 years SIS mission time at $\Delta t = 1\,\text{h}$ sums up to 1372.4 h.

7.3 Lossless state space reduction

A reasonable method for reducing the state space of standard markov models is the removal of unreachable states. Those are characterized as follows:

$$\boldsymbol{s}_i \in \mathcal{S} \text{ is unreachable, if } (\forall j \neq i)\,(\boldsymbol{A}[i][j] = 0)\,.$$

However, in SMPMMs, this criterion cannot be applied as is, as some states can only be reached via shifting operations at phase transitions. Consider, e.g., the repair state DR for passive dangerous failures. A transition to this state is only possible via events $\boldsymbol{e_{pt}}$, $\boldsymbol{e_{ppt}}$, $\boldsymbol{e_{mt}}$, and $\boldsymbol{e_{mgr}}$ from the CFSM introduced in subsec. 5.2.3 (p. 73). All of these events are encoded in the phase transition matrices, not in the transition matrix itself.

Therefore, a slightly different approach is introduced:

Definition 7.12 (State space reduction). *All state space vectors $\boldsymbol{s}_i$ can be excluded from the original state space $\mathcal{S}$ that have their index i contained in the set*

$$\mathcal{REM} = \left\{ i \mid \exists k \left(\begin{array}{l} (\lambda_{du,k} = 0 \wedge \boldsymbol{s}_i[k] \in \{DU\}) \\ \vee\, (\lambda_{dup,k} = 0 \wedge \boldsymbol{s}_i[k] \in \{DUP\}) \\ \vee\, (\lambda_{dn,k} = 0 \wedge \boldsymbol{s}_i[k] \in \{DN\}) \\ \vee\, (\lambda_{dd,k} = 0 \wedge \boldsymbol{s}_i[k] \in \{DDU\}) \\ \vee\, (\lambda_{a,k} = 0 \wedge \boldsymbol{s}_i[k] \in \{AU, AR\}) \\ \vee\, (\lambda_{du,k} = \lambda_{dup,k} = \lambda_{dn,k} = 0 \wedge \boldsymbol{s}_i[k] \in \{DR\}) \\ \vee\, (\lambda_{ddx,k} = \lambda_{dd,k} = 0 \wedge \boldsymbol{s}_i[k] \in \{DDR\}) \end{array} \right) \right\}.$$

All SMPMM objects have to be adapted by removing the i-th rows and columns from matrices and the i-th elements from vectors according to the indices contained in $\mathcal{REM}$.

The definition refers to the fact that a component which has its assigned failure rate λ_{du} set to zero can never reach its component state DU. Hence, the elements of the probability distribution vector representing state space vectors with the considered component in DU state will always be zero. If additionally λ_{du} and λ_{dn} are zero, then the DR state can also never be reached, as one of the described three failure mechanisms is required herefor. A reconsideration of the CFSM in subsec. 5.2.3 (p. 73) immediately shows the relevant scenarios considered in def. 7.12.

Example 7.5 (Demonstration of the optimization effect). Reconsider ex. 7.1 (p. 121) after the application of ASMLSIS decomposition as explained in sec. 7.2 and demonstrated in ex. 7.4. The SIS is decomposed into three subsystems with 4 - 2 - 4 contained components each. Without loss of generality the final element part can be considered to always not have any *dd* failures. The sensor part usually has a proof test coverage of 100% and thus $\lambda_{dn} = 0$. Additionally, partial proof tests are typically applied to the final element part only such that the DUP state cannot be reached for sensor components and the logic solver. This minimal a priori information enables for recalculating memory consumption of the transition matrix and the calculation time for a 30 years mission time of the SIS.

The transition matrix size reduces to $7^{4 \cdot 2}$ - $8^{2 \cdot 2}$ - $8^{4 \cdot 2}$ elements with a memory consumption of 46.1 - 0.03 - 134.2, i.e., 180.3 megabytes in total. The calculation time for a single matrix multiplication drops to $4.6 \cdot 10^{-1}$ - $8.7 \cdot 10^{-6}$ - 2.3, i.e., 2.76 s. With that, the calculation time for a 30 years SIS mission time at $\Delta t = 1\,\mathrm{h}$ sums up to 201.5 h.

Notice that this example performs a state space reduction based on typical expectations only. Realistic systems usually allow for much higher reduction.

7.4 Sparse matrix implementation

Transition matrices and phase transition matrices are sparse matrices [Rau04]. Only a few system states can be reached from any given state $s_i \in \mathcal{S}$ via transition. Most important reason is the single failure principle. During a discrete time step Δt only one component failure may occur. This results in the impossibility of having a state with one failed component connected to any state with more than two failed components, rendering the largest part of the state space unreachable via single transition from a given state. Typical transition matrices have an occupancy of around 10%. More remarkably, phase transition matrices have a single 1-entry per column, the rest are zeros (compare the related def. 6.16 (p. 117)). This leads to Q non-zero entries in a $Q \times Q$ matrix, i.e., an occupancy of $\frac{1}{Q}$ which drops below 10% for $Q > 10$ and lies around $1.5 \cdot 10^{-4}$ for the state space related to four components without state space reduction according to sec. 7.3 (p. 133). Hence, a sparse matrix implementation enables for a significant performance optimization [Rau04].

Basic principle is to store matrices not in arrays of size $Q \times Q$, but to use double linked list structures which hold one matrix element together with its indices each. At first glance this principle seems to be impractical, since an overhead for the index variables and two forward pointers is required for each non-zero matrix element. But the larger a matrix is and the more sparse it is occupied, the better this implementation performs. For reasons of simplicity it is assumed that the element value variable as well as the index variables and the two required forward pointers all require an equal amount of memory m. A single standard matrix element thus requires m memory units, and a sparse matrix implementation (SMI) element $5m$. The matrix occupancy may be denoted with $o \in [0, 1]$, such that $o \cdot Q^2$ is the number of non-zero elements in $\boldsymbol{A}$. Then the memory consumption functions for standard implementation (SI) and SMI for a $Q \times Q$ matrix can be denoted as

$$
\begin{aligned}
\mathrm{Mem}_{SI}(Q) &= Q^2 m, \\
\mathrm{Mem}_{SMI}(Q, o) &= Q^2 5 m o.
\end{aligned}
$$

Relating both equations delivers

$$
\begin{aligned}
\mathrm{Mem}_{SI}(Q) &> \mathrm{Mem}_{SMI}(Q, o) \\
1 > 5o &\Rightarrow o < \frac{1}{5}.
\end{aligned}
$$

For all matrices with occupancy lower than 20%, the SMI approach results in a lower memory consumption than the SI.

The number of required multiplications can also be significantly reduced, as only multiplications $\boldsymbol{A}[i_1][j_1] \cdot \boldsymbol{A}[i_2][j_2]$ with $\boldsymbol{A}[i_1][j_1] \neq 0$ and $\boldsymbol{A}[i_2][j_2] \neq 0$ are executed. The required pointer operations can be neglected, as no floating point multiplications are involved.

The worst-case, i.e., the way to arrange the $o \cdot Q^2$ matrix elements in $\boldsymbol{A}$ such that the maximum number of multiplications follows, is a square-sized submatrix $\boldsymbol{A}'$ inside the transition matrix with $\boldsymbol{A}'[1][1]$ on the diagonal of $\boldsymbol{A}$.

$$\boldsymbol{A} = \begin{pmatrix} 1 & 1 & 1 & 0 & 0 \\ 1 & 1 & 1 & 0 & 0 \\ 1 & 1 & 1 & 0 & 0 \\ 0 & 0 & 0 & 0 & 0 \\ 0 & 0 & 0 & 0 & 0 \end{pmatrix} \tag{7.1}$$

In eq. 7.1, an exemplary $Q \times Q = 5 \times 5$ matrix $\boldsymbol{A}$ with $o = \frac{9}{25}$ is denoted. The required number of multiplications can be calculated as q^3, where $q = 3$ is the size of the square-sized submatrix $\boldsymbol{A}'$. If o is given such that the related number of non-zero matrix elements has an integer square root, then the required number of multiplications can be generally derived as $\left(\sqrt{Q^2 o}\right)^3 = Q^3 o^{\frac{3}{2}}$. $\boldsymbol{A}'$ hereby is a $\sqrt{Q^2 o} \times \sqrt{Q^2 o}$ matrix. The computation time functions for SI and SMI approach are

$$\begin{aligned} \mathrm{Com}_{SI}(Q) &= Q^3 t, \text{ and} \\ \mathrm{Com}_{SMI}(Q, o) &= Q^3 o^{\frac{3}{2}} t, \end{aligned}$$

with t as the calculation time for a single multiplication. The relation of both functions delivers

$$\begin{aligned} \mathrm{Com}_{SI}(Q) &> \mathrm{Com}_{SMI}(Q, o) \\ Q^3 t > Q^3 o^{\frac{3}{2}} t &\Rightarrow 1 > o^{\frac{3}{2}} \Rightarrow o < 1, \end{aligned}$$

i.e., for any occupation o below 100% an increase in computation speed is achieved. The improvement factor is

$$\frac{\mathrm{Com}_{SMI}(Q, o)}{\mathrm{Com}_{SI}(Q)} = o^{\frac{3}{2}}.$$

Even with this worst case assumption, it is obvious that a significant optimization in memory consumption and computation time is possible for nearly all possible SISs. Notice that for non-integer square roots $\sqrt{Q^2 o}$ the submatrix $\boldsymbol{A}'$ cannot be constructed properly. In that case $\mathrm{floor}(\sqrt{Q^2 o}) + 1$ could be chosen as the dimension of $\boldsymbol{A}'$, resulting in another conservatism ('floor' is the round down function). But as this does not significantly impact on the abovestanding results, it is not considered in the closing example.

Example 7.6 (Demonstration of the optimization effect). The decomposed SIS with 4 - 2 - 4 components in its respective subsystems (ex. 7.4 (p. 132)) and reduced state space (ex. 7.5 (p. 133)) is now supported by a sparse matrix implementation. The dimension of the three original transition matrices is $Q_1 = 7^4$, $Q_2 = 8^2$, and $Q_3 = 8^4$. A conservative occupancy of 15% is assumed from now on. Then the required memory consumption with $m = 8$ bytes is

$$\begin{aligned}
\text{Mem}_{SMI}(7^4, 0.15) &= 7^{4\cdot 2} \cdot 5 \cdot 8 \cdot 0.15\text{bytes} \approx 34.6 \text{ megabytes},\\
\text{Mem}_{SMI}(8^2, 0.15) &= 8^{2\cdot 2} \cdot 5 \cdot 8 \cdot 0.15\text{bytes} \approx 2.5 \cdot 10^{-2} \text{ megabytes},\\
\text{Mem}_{SMI}(8^4, 0.15) &= 8^{4\cdot 2} \cdot 5 \cdot 8 \cdot 0.15\text{bytes} \approx 100.7 \text{ megabytes},
\end{aligned}$$

resulting in a total memory consumption of 135.3 megabytes. The related computing times are

$$\begin{aligned}
\text{Com}_{SMI}(7^4, 0.15) &= 7^{4\cdot 3} 0.15^{\frac{3}{2}} \frac{1}{30 \cdot 10^9} \,\text{s} \approx 2.7 \cdot 10^{-2}\,\text{s},\\
\text{Com}_{SMI}(8^2, 0.15) &= 8^{2\cdot 3} 0.15^{\frac{3}{2}} \frac{1}{30 \cdot 10^9} \,\text{s} \approx 5.1 \cdot 10^{-7}\,\text{s},\\
\text{Com}_{SMI}(8^4, 0.15) &= 8^{4\cdot 3} 0.15^{\frac{3}{2}} \frac{1}{30 \cdot 10^9} \,\text{s} \approx 1.3 \cdot 10^{-1}\,\text{s},
\end{aligned}$$

i.e., $1.6 \cdot 10^{-1}$ s in total. With that, the calculation time for a 30 years SIS mission time at $\Delta t = 1$ h sums up to 11.7 h. This result, based on three simple optimization approaches delivers the desired results in a reasonable amount of time while consuming a realistic amount of memory. A system as described in this continued example can be calculated on a modern standard PC.

7.5 Further optimzation approaches

7.5.1 Component aggregation

The international safety standards such as IEC 61511 and IEC 61508 do not differentiate between component instances in a channel. The provided calculation formulas assume that the utilized failure rates represent complete channels. Common practice for safety engineers is to collect all component individual failure rates and sum them up in order to retrieve the channel failure rate. This obviously results in a loss of information compared to the ASML approach as presented in this thesis, since, e.g., the inhibition effect gets entirely lost, and cannot be reproduced. However, merging multiple components into a single representative 'channel component' reduces the related SMPMM's state space significantly and is therefore worthwhile an implementation, especially as it is easy to formalize.

Definition 7.13 (Component aggregation). *Two arbitrary components c_i and c_j can be merged if*

1. $\exists\,(s_k \in \mathcal{SEQ})\,\exists m \exists n\,(m + 1 = n \wedge \boldsymbol{iv}_k[m] = c_i \wedge \boldsymbol{iv}_k[n] = c_j)$, *i.e., if exists a sequence s_k with components c_i and c_j as input vertices with adjacent indices.*

2. $\mathcal{PT}_i = \mathcal{PT}_j$
3. $\mathcal{PPT}_i = \mathcal{PPT}_j$
4. $\mathcal{MT}_i = \mathcal{MT}_j$
5. $\boldsymbol{\mu}_i = \boldsymbol{\mu}_j$

Then an aggregated component c_a can be constructed according to

$$c_a = (\boldsymbol{\lambda}_a, \boldsymbol{\mu}_a, \sigma_a, \mathcal{PT}_a, \mathcal{PPT}_a, \mathcal{MT}_a),$$

where $\boldsymbol{\mu}_a = \boldsymbol{\mu}_i$, $\mathcal{PT}_a = \mathcal{PT}_i$, $\mathcal{PPT}_a = \mathcal{PPT}_i$, $\mathcal{MT}_a = \mathcal{MT}_i$ and $\forall l \, (\boldsymbol{\lambda}_a[l] = \boldsymbol{\lambda}_i[l] + \boldsymbol{\lambda}_j[l])$. The set of components is updated according to

$$\mathcal{COM} = (\mathcal{COM} \cup \{c_a\}) \setminus \{c_i, c_j\},$$

i.e., c_i and c_j are removed from the set of components and the new aggregated component c_a is inserted. Sequence s_k is updated with a new input vertex vector $\boldsymbol{iv}'_k$ according to

$$\begin{aligned}
&|\boldsymbol{iv}'_k| = |\boldsymbol{iv}_k| - 1, \\
&\forall (l < m) \, (\boldsymbol{iv}'_k[l] = \boldsymbol{iv}_k[l]), \\
&\boldsymbol{iv}_k[m] = c_a, \\
&\forall (l > n) \, (\boldsymbol{iv}'_k[l-1] = \boldsymbol{iv}_k[l]).
\end{aligned}$$

c_i and c_j are removed and c_a is inserted.

This definition seems to be very rigid at first glance due to requirements 1 to 4. On the other hand, the component individual proof test times and replacement times are an important feature of the ASML approach. Most other calculation approaches do not allow for component individual parameterization and therefore realistic channel components often have equal parameterization anyway. Hence, existing SISs have a large potential for successfully applying the component aggregation approach.

Example 7.7 (Demonstration of the optimization effect)**.** The decomposed SIS with 4 - 2 - 4 components in its respective subsystems (ex. 7.4 (p. 132)), reduced state space (ex. 7.5 (p. 133)), as well as sparse matrix implementation (ex. 7.6 (p. 135)) is now optimized by channel component aggregation. If the described SIS has two sensor channels and two final element channels consisting of two components each, then the transition matrices are of dimension $Q_1 = 7^2$, $Q_2 = 8^2$, and $Q_3 = 8^2$. With a conservative occupancy of 15%, the required memory consumption with $m = 8$ bytes is

$$\begin{aligned}
\mathrm{Mem}_{SMI}(7^2, 0.15) &= 7^{2 \cdot 2} \cdot 5 \cdot 8 \cdot 0.15 \text{bytes} \approx 1.4 \cdot 10^{-2} \text{ megabytes}, \\
\mathrm{Mem}_{SMI}(8^2, 0.15) &= 8^{2 \cdot 2} \cdot 5 \cdot 8 \cdot 0.15 \text{bytes} \approx 2.5 \cdot 10^{-2} \text{ megabytes}, \\
\mathrm{Mem}_{SMI}(8^2, 0.15) &= 8^{2 \cdot 2} \cdot 5 \cdot 8 \cdot 0.15 \text{bytes} \approx 2.5 \cdot 10^{-2} \text{ megabytes},
\end{aligned}$$

resulting in a total memory consumption of $6.4 \cdot 10^{-2}$ megabytes. The related computing times are

$$\begin{aligned}
\mathrm{Com}_{SMI}(7^2, 0.15) &= 7^{2\cdot 3} 0.15^{\frac{3}{2}} \frac{1}{30 \cdot 10^9}\,\mathrm{s} \approx 2.3 \cdot 10^{-7}\,\mathrm{s},\\
\mathrm{Com}_{SMI}(8^2, 0.15) &= 8^{2\cdot 3} 0.15^{\frac{3}{2}} \frac{1}{30 \cdot 10^9}\,\mathrm{s} \approx 5.1 \cdot 10^{-7}\,\mathrm{s},\\
\mathrm{Com}_{SMI}(8^2, 0.15) &= 8^{2\cdot 3} 0.15^{\frac{3}{2}} \frac{1}{30 \cdot 10^9}\,\mathrm{s} \approx 5.1 \cdot 10^{-7}\,\mathrm{s},
\end{aligned}$$

i.e., $1.3 \cdot 10^{-6}\,\mathrm{s}$ in total. With that, the calculation time for a 30 years SIS mission time at $\Delta t = 1\,\mathrm{h}$ sums up to 0.34 s. If the slight inaccuracy of the results is accepted, then the component aggregation provides a significant reduction of the state space, as it aims at decreasing the exponent of the basic state space size formula $|\mathcal{CS}|^{N_{SIS}}$.

7.5.2 State lumping

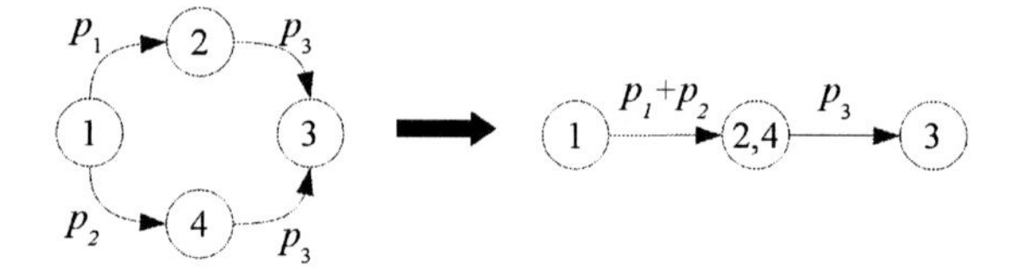

Figure 7.6: Exemplary state space reduction by state lumping

A common method to reduce the state space of an MM is to conduct an investigation on state lumpability. The basic idea is to create a new MM which has multiple states of the original MM condensed into a single representative state. In literature, a variety of lumping approaches have been introduced allowing for various degrees of accuracy (see, e.g., [Sie95] for a good overview). With respect to this thesis, the following lumping rule is easily applicable:

Definition 7.14 (Lumpability). *Two states $\boldsymbol{s_i}$ and $\boldsymbol{s_j}$ of an SMPMM can be lumped together into an aggregated state $\boldsymbol{s_k}$ if*

- *they have identical sets of follow-up states*
- *their respective transitions to the follow-up states have identical probabilities*
- *they are identical with respect to their contribution to the desired unavailability characteristics*

Then all transitions from arbitrary states to either $\boldsymbol{s_i}$ or $\boldsymbol{s_j}$ are redirected to $\boldsymbol{s_k}$. If thereby multiple transitions from a single state to $\boldsymbol{s_k}$ occur, then these transitions are merged together by summing up their probabilities. The transitions to the follow-up states of either $\boldsymbol{s_i}$ or $\boldsymbol{s_j}$ (they are identical) are applied to the aggregated state $\boldsymbol{s_k}$.

Some remarks on def. 7.14:

- 'Follow-up states' to any state $\boldsymbol{s}_m$ are all states $\{\boldsymbol{s}_n \mid \boldsymbol{A}[n][m] \neq 0\}$.
- ' identical with respect to their contribution to the desired unavailability characteristics' means that for the considered states $\boldsymbol{s}_i$ and $\boldsymbol{s}_j$ $\boldsymbol{c}_{PFD}[i] = \boldsymbol{c}_{PFD}[j]$ and $\boldsymbol{c}_{PFS}[i] = \boldsymbol{c}_{PFS}[j]$ holds.

Example 7.8 (Application of the state lumping rule)**.** See fig. 7.6 for an exemplary MM which encourages the application of def. 7.14. The sets of follow-up states for all states $\boldsymbol{s}_1$ to $\boldsymbol{s}_4$ are

$$\mathcal{FUS}_{\boldsymbol{s}_1} = \{\boldsymbol{s}_2, \boldsymbol{s}_4\}, \mathcal{FUS}_{\boldsymbol{s}_2} = \{\boldsymbol{s}_3\}, \mathcal{FUS}_{\boldsymbol{s}_3} = \{\}, \mathcal{FUS}_{\boldsymbol{s}_4} = \{\boldsymbol{s}_3\}.$$

Only $\boldsymbol{s}_2$ and $\boldsymbol{s}_4$ have identical sets of follow-up states and thus serve as the lumping candidates. As for both states

$$\mathrm{P}\left(\boldsymbol{s}((k+1)\Delta t) = \boldsymbol{s}_3 \mid \boldsymbol{s}(k\Delta t) = \boldsymbol{s}_2\right) = \mathrm{P}\left(\boldsymbol{s}((k+1)\Delta t) = \boldsymbol{s}_3 \mid \boldsymbol{s}(k\Delta t) = \boldsymbol{s}_4\right) = p_3$$

holds, merging states $\boldsymbol{s}_i$ and $\boldsymbol{s}_j$ is possible (assuming that both represent either an s-available or s-unavailable system).

This approach clearly does not provide the potential of the beforementioned optimization techniques, as the potential of state merging is supposed to be significantly less than a whole order of magnitude. In MPMMs additional difficulties arise, as it is to be investigated whether lumping has an impact on shifting procedures. However, although currently not implemented, further investigations on the field of state lumping techniques is recommended.

Chapter 8

Validation

8.1 Overview

Validation of the results generated with the ASML approach is not straight forward. Difficulties arise from the lack of certified benchmark results for complex SIS behavior. Most available frameworks for safety calculation are either expert-systems and require extended engineering skills in order to model the required systems (e.g., [PR99]) or too simple to be capable of reproducing the desired effects (e.g., [DIN02], [GY08]). The former provide such a high degree of freedom that an uncertainty arises about whether particular SIS effects have been implemented correctly or not. Even simple approaches such as a monte carlo simulation lead to large complexity in the implementation and modeling phase. Therefore, the related results can not serve as a valid benchmark, as their correctness is not guaranteed. The latter suffice for basic validation tasks and are, among others, referred to in the course of this chapter.

Various methods are consulted in order to provide a sufficient amount of validation for the ASML approach. Numerous examples for each method are provided, using concrete numerical instrumentation scenarios. The considered methods are

1. Investigation of specific ASML configurations with a priori knowledge about the calculation results, i.e., the expected behavior: sec. 8.2.

2. Cross-check with as many qualified sources as possible. This includes a comparison of ASML results with tabular PFD values from IEC 61508 for basic scenarios: sec. 8.3.

3. Cross-check with additional sources providing extensions to the standard's (legitimate) formulas such that unavailabilities for, e.g., diversely instrumented SISs, can be numerically compared: sec. 8.4.

4. Investigation of the impact of ASML-specific, non-standard effects on the overall system unavailability: sec. 8.5.

In the following subsections the component specific characteristics are always chosen as $\forall j\left(\lambda_i[j] = 0\,\mathrm{h}^{-1}\right)$, $\forall k\left(\mu_i[k] = 0\,\mathrm{h}^{-1}\right)$, $\mathcal{PT} = \mathcal{PPT} = \mathcal{MT} = \emptyset$, if not specified otherwise.

Accordingly, $\mathcal{MGS} = \emptyset$. All denoted visualizations of transition matrices (TMVs) are generically created using the ASML approach. The same holds for all calculation results obtained from ASML descriptions. In all TMVs, dangerous states, i.e., PFD relevant states, are labeled with a double line.

8.2 A priori knowledge validation

8.2.1 Unreliability

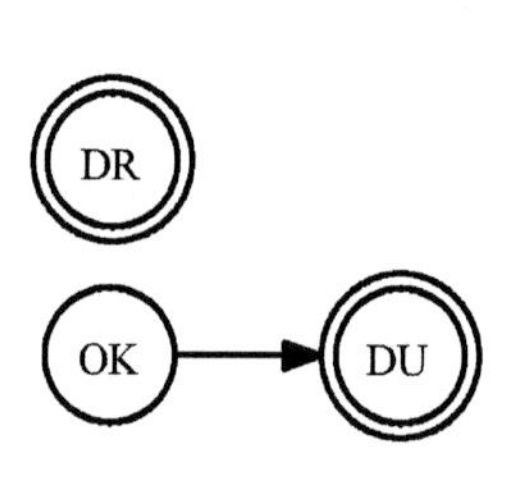

Figure 8.1: TMV for a pure RF process

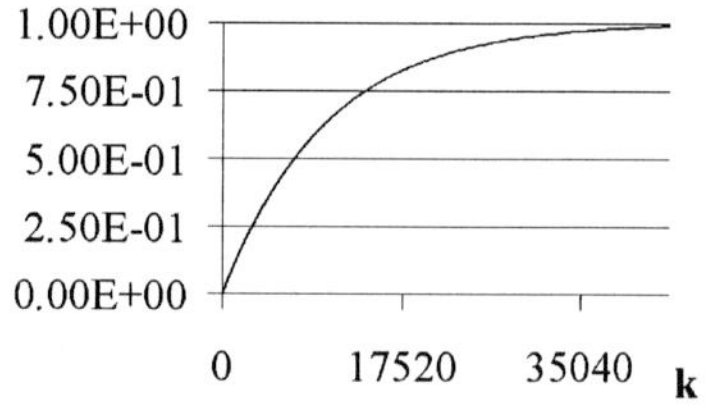

Figure 8.2: PFD over discrete time ($\Delta t = 1\,\mathrm{h}$) for a pure RF process

A single component without any kind of maintenance is subject to RF processes only, i.e., is characterized by its unreliability. This can be replicated using a single ASML component c_1 with $\lambda_{du,1} = 100,000\,\mathrm{FIT}$ and empty sets $\mathcal{PT}$ and $\mathcal{MT}$. Any *du* failure occurrence will thus never get revealed since no proof tests or replacements are conducted. The PFD over time for a SIS mission time $\tau_{mt} = 5\,\mathrm{y}$ is denoted in fig. 8.2. The behavior is as expected: as introduced in sec. 3.5.5 (p. 40), the unavailability of an unmaintained SIS is equal to its unreliability. The curve corresponds to that of a pure RF process with exponentially distributed lifetime (see sec. 3.3) with $\lambda = \lambda_{du,1}$. The related TMV (fig. 8.1) is equal to the one shown in fig. 4.10 (p. 53) except for the additional repair state *DR*. It is not connected to the rest of the graph, as it can only be reached via phase transitions and can be left via repair transitions only. The related state probability is therefore $\mathrm{P}(s_{DR}) = 0 \forall k$. This implies that the state probability $\mathrm{P}(s_{DU})(k)$ alone represents the PFD. The set of phase transitions contains no elements, i.e., $\mathcal{T} = \emptyset$. The generated SMPMM has the form of the simple MM introduced in subsec. 4.4.2 and is depicted in fig. 4.10 (p. 53).

8.2.2 Unavailability: RFR-DRI type C cycle

Adding maintenance changes the resulting PFD curve fundamentally. The RF cycle from the previous subsection transforms into a type C RFR-DRI cycle (see subsec. 3.5.4 (p. 40)) in case of periodically applied proof tests. With $\lambda_{du,1} = 100,000\,\mathrm{FIT}$ again, $\mu_{dr,1} = 0.125\,\mathrm{h}^{-1}$, $\mathcal{PT}_1 = \{1\,\mathrm{y}, 2\,\mathrm{y}, 3\,\mathrm{y}, 4\,\mathrm{y}\}$ and $\tau_{mt} = 5\,\mathrm{y}$, the resulting PFD over time is

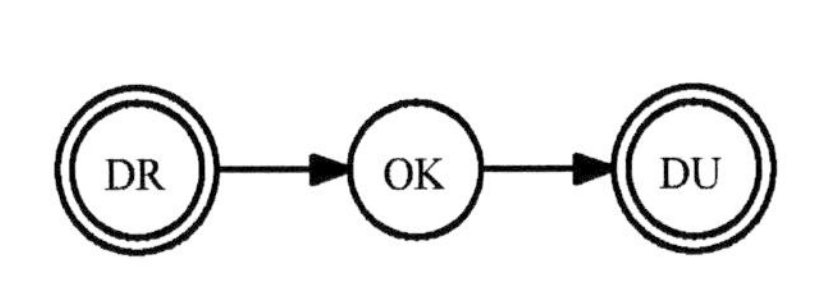

Figure 8.3: TMV for an RFR-DRI type C cycle

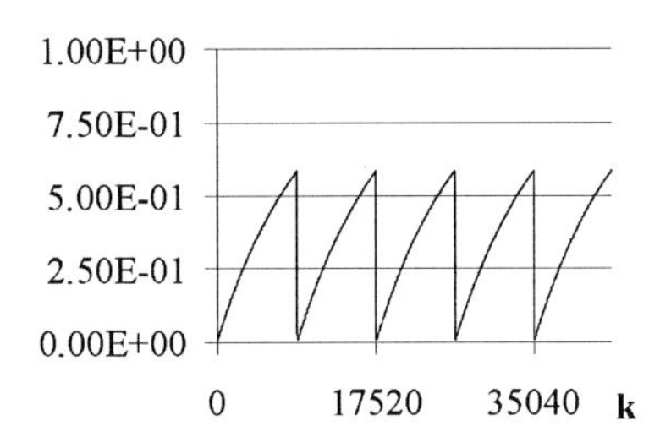

Figure 8.4: PFD over discrete time (Δt = 1 h) for an RFR-DRI type C cycle

a sawtooth curve (fig. 8.4). In contrast to the previous case, the repair state DR can be reached via phase transition and left with a valid repair rate $\mu_{dr,1}$ (see TMV in fig. 8.3). With state space $\mathcal{S} = \{(OK)^\top, (DU)^\top, (DR)^\top\}$, a single phase transition matrix $\mathcal{M}_1$ matrix as specified in def. 6.17 (p. 118) is generated according to

$$\boldsymbol{M}_1 = \begin{bmatrix} 1 & 0 & 0 \\ 0 & 0 & 0 \\ 0 & 1 & 1 \end{bmatrix},$$

i.e., at phase transitions the state probability of $\boldsymbol{s}_{DU}$ is shifted to $\boldsymbol{s}_{DR}$. The related phase transition set is $\mathcal{T} = \{(1\,\text{y}, \boldsymbol{M}_1), (2\,\text{y}, \boldsymbol{M}_1), (3\,\text{y}, \boldsymbol{M}_1), (4\,\text{y}, \boldsymbol{M}_1)\}$. The generated SMPMM has the form of the simple MM introduced in subsec. 4.4.5 and is depicted in fig. 4.15 (p. 57).

8.2.3 Unavailability: RFR-IRI Cycle

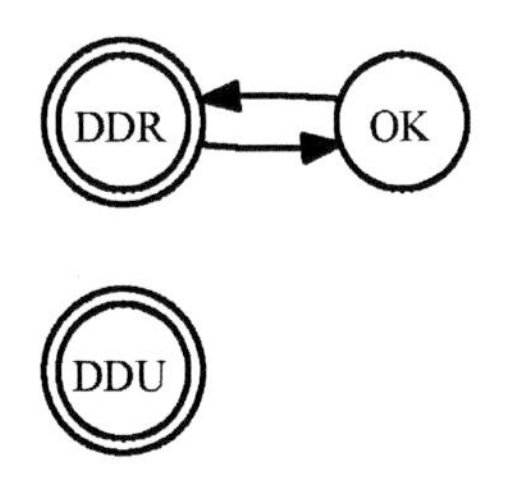

Figure 8.5: TMV for an RFR-IRI cycle

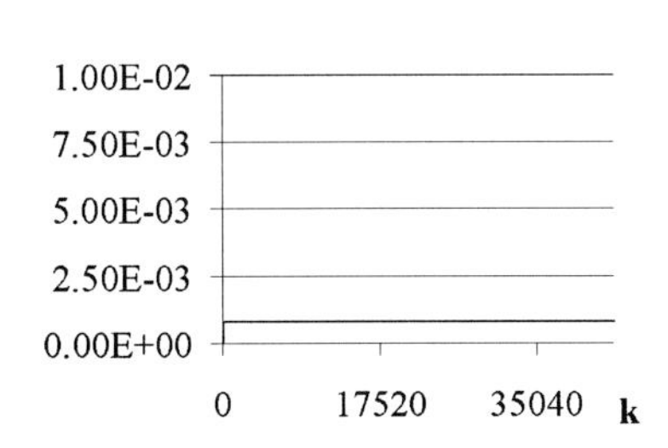

Figure 8.6: PFD over discrete time (Δt = 1 h) for an RFR-IRI cycle

With $\lambda_{dd,1} = 100,000\,\text{FIT}$ and $\mu_{ddr,1} = 0.125\,\text{h}^{-1}$, an RFR-IRI process according to subsec. 3.5.1 (p. 36) is generated. The related PFD over time for $\tau_{mt} = 5\,\text{y}$ is depicted in fig. 8.6. The steady state value for the PFD is reached quickly and can be calculated as

$$\frac{\lambda_{dd,1}}{\lambda_{dd,1} + \mu_{ddr,1}} = \frac{10^{-4}\,\text{h}^{-1}}{10^{-4}\,\text{h}^{-1} + 0.125\,\text{h}^{-1}} \approx 8.0 \cdot 10^{-4}$$

(see eq. 3.11 (p. 36)). The calculation results therefore perfectly meet the expectations. The TMV in fig. 8.5 denotes states OK and DDR mutually connected by transitions. This configuration has been introduced for simple MMs in subsec. 3.5.1 and is depicted in fig. 4.11 (p. 54). The additional DDU state is unreachable, as inhibition effects for the single component cannot occur. This implies that the state probability $\text{P}(\boldsymbol{s}_{DDR})(k)$ alone represents the PFD.

8.2.4 Unavailability: multicomponent system

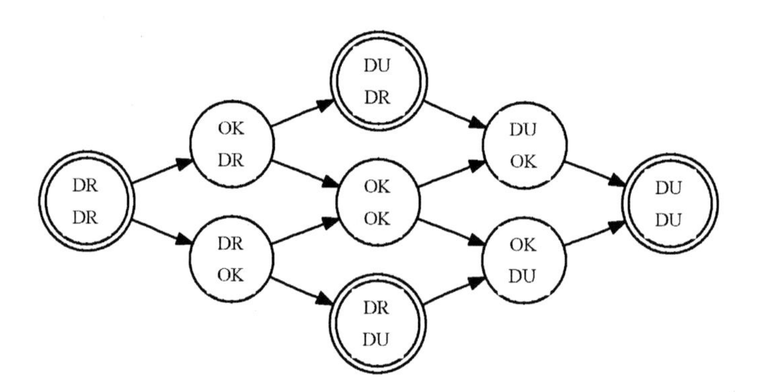

Figure 8.7: TMV for a multi component system

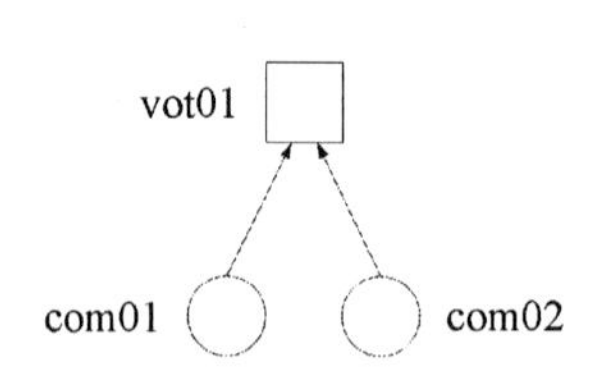

Figure 8.8: ASMLG for a simple multi component SIS

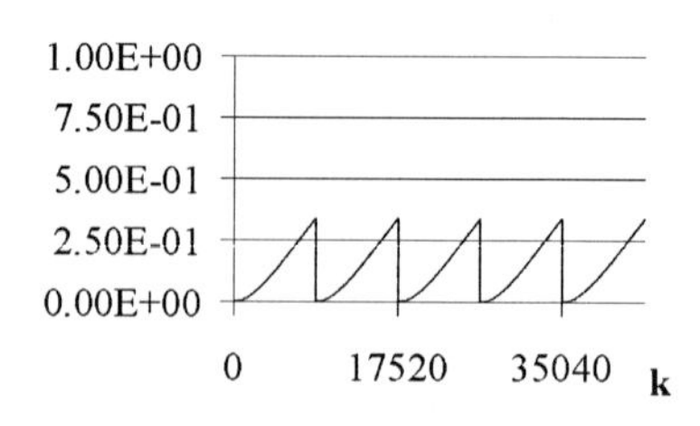

Figure 8.9: PFD over discrete time ($\Delta t = 1\,\text{h}$) for the multi component SIS

A double component system with global periodic proof tests has been basically introduced in subsec. 4.4.6 (p. 57) with a related TMV depicted in fig. 4.16 (p. 58). Consider the ASMLG in fig. 8.8. The voter vot01 is configured as a static 1oo2 redundancy. Both components are parameterized with $\lambda_{du,\text{com1}} = \lambda_{du,\text{com2}} = 100,000\,\text{FIT}$ and respective repair rates $\mu_{dr,\text{com1}} = \mu_{dr,\text{com2}} = 0.125\,\text{h}^{-1}$ as well as $\mathcal{PT}_{\text{com1}} = \mathcal{PT}_{\text{com2}} = \{1\,\text{y}, 2\,\text{y}, 3\,\text{y}, 4\,\text{y}\}$. The ASML approach delivers a SMPMM with TMV according to fig. 8.7. As the generator equations always construct type C RFR-DRI cycles, additional repair states occur with one or both components in DR state. For a mission time $\tau_{mt} = 5\,\text{y}$, a PFD according to fig. 8.9 results. Since the components are stochastically independent from each other, i.e., no maintenance groups and inhibition effects impact on the result, the SIS unavailability can be retrieved as $\text{U}_{SIS}(k) = \text{U}_{\text{com1}}(k) \cdot \text{U}_{\text{com2}}(k)$. The depicted PFD curve lives up to these expectations when comparing fig. 8.4 for the single component with fig. 8.9 for the double component case. As both components are simultaneously proof tested, a single phase transition matrix

$$\boldsymbol{M}_1 = \begin{bmatrix} 1 & 0 & 0 & 0 & 0 & 0 & 0 & 0 & 0 \\ 0 & 0 & 0 & 0 & 0 & 0 & 0 & 0 & 0 \\ 0 & 1 & 1 & 0 & 0 & 0 & 0 & 0 & 0 \\ 0 & 0 & 0 & 0 & 0 & 0 & 0 & 0 & 0 \\ 0 & 0 & 0 & 0 & 0 & 0 & 0 & 0 & 0 \\ 0 & 0 & 0 & 0 & 0 & 0 & 0 & 0 & 0 \\ 0 & 0 & 0 & 1 & 0 & 0 & 0 & 0 & 0 \\ 0 & 0 & 0 & 0 & 0 & 0 & 0 & 0 & 0 \\ 0 & 0 & 0 & 0 & 1 & 1 & 0 & 1 & 1 \end{bmatrix}$$

is sufficient to describe all reset processes for the state space

$$\mathcal{S} = \left\{ \begin{array}{l} (OK\;OK)^\top, (OK\;DU)^\top, (OK\;DR)^\top, (DU\;OK)^\top, \\ (DU\;DU)^\top, (DU\;DR)^\top, (DR\;OK)^\top, (DR\;DU)^\top, (DR\;DR)^\top \end{array} \right\}.$$

The phase transition matrix encodes, e.g., a shifting process $\boldsymbol{s}_2 \rightarrow \boldsymbol{s}_3$, i.e., $(OK\;DU)^\top \rightarrow (OK\;DR)^\top$ via $\boldsymbol{M}_1[3][2] = 1$. The related phase transition set is $\mathcal{T} = \{\,(1\,\text{y}, \boldsymbol{M}_1), (2\,\text{y}, \boldsymbol{M}_1), (3\,\text{y}, \boldsymbol{M}_1), (4\,\text{y}, \boldsymbol{M}_1)\,\}$.

8.2.5 Signal replication

Figure 8.10 depicts two SISs. SIS_1 consists of a single component com01 and a static 1oo2 voter vot01 with an input vertex vector of length 2. Both input vertices are specified as com01, i.e., the component's signal is replicated and connected to multiple parts of the SIS. SIS_1 obviously does not contain operative redundancy even though the 1oo2 voter suggests so. A similar system has already been discussed in subsec. 4.6.1 (p. 66) with related fig. 4.23 (p. 67). The expected unavailability equals the results of the reference system SIS_2. Here, component com02 feeds a static 1oo1 voter vot02. With $\lambda_{du,\text{com01}} = \lambda_{du,\text{com02}} = 100,000\,\text{FIT}$, $\mu_{dr,\text{com01}} = \mu_{dr,\text{com02}} = 0.125\,\text{h}^{-1}$, $\mathcal{PT}_{\text{com01}} = \mathcal{PT}_{\text{com02}} = \{1\,\text{y}, 2\,\text{y}, 3\,\text{y}, 4\,\text{y}\}$

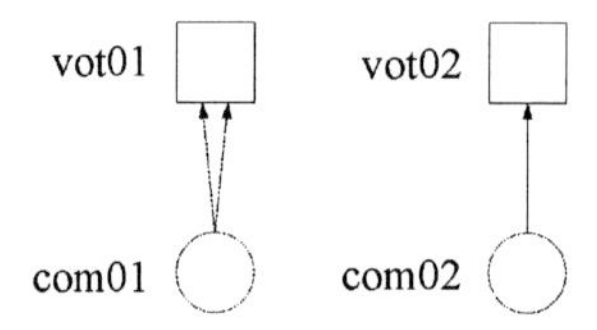

Figure 8.10: ASMLGs of SIS_1 and SIS_2 in the signal replication example

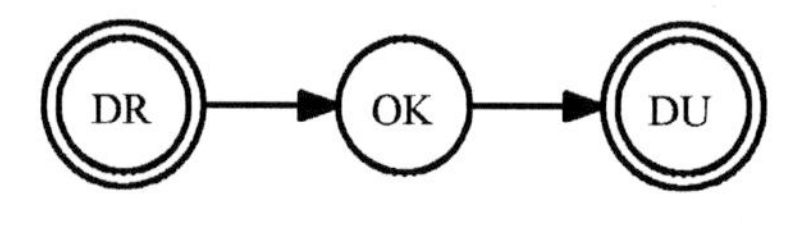

Figure 8.11: TMVs for SIS_1 and SIS_2 in the signal replication example

and a mission time $\tau_{mt,1} = \tau_{mt,2} = 5\,\mathrm{y}$, the respective SMPMMs can be generated. Both ASMLGs lead to the exact same SMPMM for which the TMV is depicted in fig. 8.11. The related PFD over discrete time is already known from fig. 8.4. Obviously, SIS_1 and SIS_2 describe both nothing but a standard type C RFR-DRI cycle in a more or less complicated way. Not even the 1oo1 voter vot02 in SIS_2 is required, as for the given parameterization no specific feature of the voter is used. Therefore, the PFD of SIS_2 is the same as for a SIS consisting of a single component only (compare to subsec. 8.2.2).

8.2.6 Ambiguity of ASMLGs

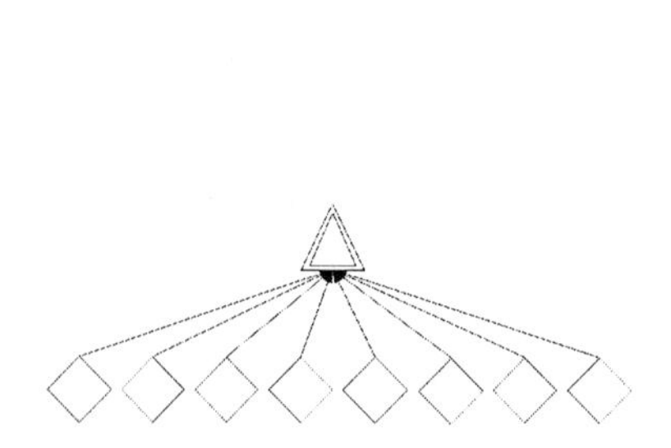

Figure 8.12: DASMLG generated from the left ASMLG in fig. 5.5

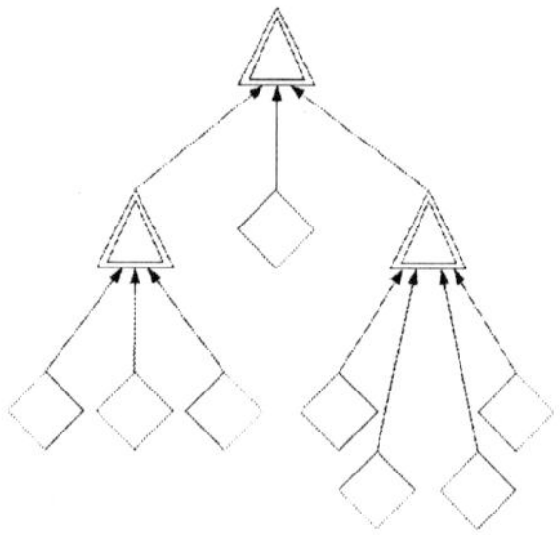

Figure 8.13: DASMLG generated from the right ASMLG in fig. 5.5

The introductory chapter 5 about ASML pointed out that ASMLGs are ambiguous, i.e., identical behavior can be achieved with differing ASMLGs. A glance of this effect has already been shown in the previous subsection. Reconsider the ASMLGs in fig. 5.5 (p. 80). Choosing the respective component failure rates in both graphs as $\lambda_{du,i} = 10,000\,\mathrm{FIT}$ and the repair rates according to $\mu_{dr,i} = 0.125\,\mathrm{h}^{-1}$, setting $\mathcal{PT}_i = \{1\,\mathrm{y}, 2\,\mathrm{y}, 3\,\mathrm{y}, 4\,\mathrm{y}\}$ and $\tau_{mt} = 5\,\mathrm{y}$, then the calculation for the desired target characteristics can be conducted. As both

ASMLGs have at least one sequence as their respective system output, a decomposition of the ASMLSISs into related DASMLSISs is possible. The resulting graphs are depicted in fig. 8.12 and fig. 8.13, respectively. As the original ASMLGs consist of components and sequences only, a decomposition down to subsystems each representing a single component is possible.

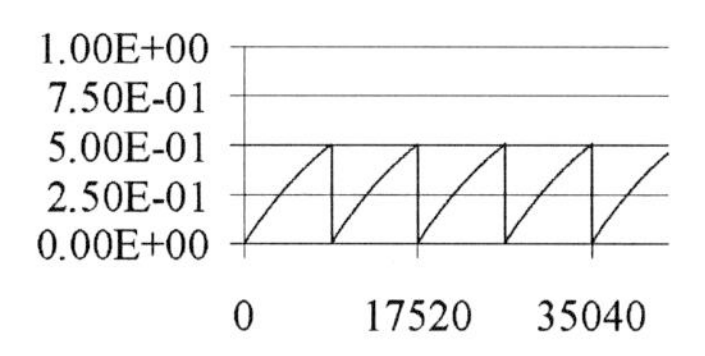

Figure 8.14: PFD over discrete time ($\Delta t = 1\,\text{h}$)

The resulting PFD over discrete time is denoted in fig. 8.14. It is identical for both SISs, just as expected. A remarkable detail is the fact, that the second variant as depicted in fig. 8.13 has a calculation time of around 20 % lower than the first one. The reason for that is the explosion of combinatorial terms for the combinatorial sequence with eight input vertices in fig. 8.12. The SMPMMs denoted by the respective subsystems are all equal to the SMPMM represented by the TMV in fig. 8.3 (p. 143). Notice that a calculation without transformation to a DASMLSIS is of course possible. The related transition matrix is of dimension 6561×6561 with 34992 non-zero elements, resulting in an occupancy around $8.1 \cdot 10^{-4}$. The calculation time on a dual core machine at 1.66GHz lies around 2.5 h which is 1700 times slower than the decomposed variant.

8.3 Qualified source validation

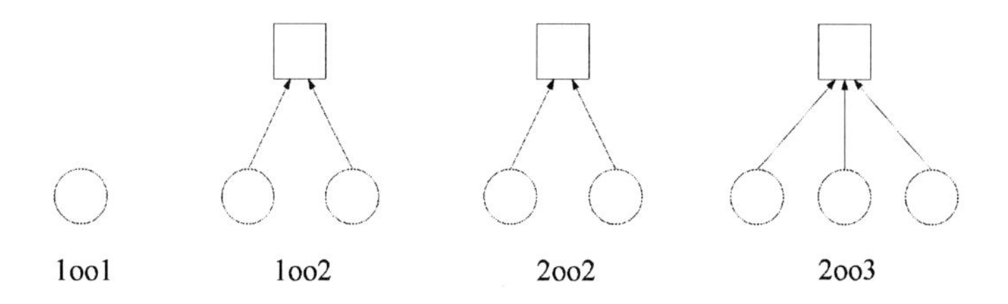

Figure 8.15: ASMLGs representing IEC 61508 system parts

IEC 61508 provides calculation formulas for 1oo1, 1oo2, 2oo2 and 2oo3 system parts. The underlying channel oriented point of view can be replicated with the ASML approach by using ASMLGs according to fig. 8.15. The static voters are configured as follows:

$$1oo2 : \mathcal{VS}_1 = \mathcal{VS}_2 = \mathcal{VS}_3 = \left\{ (1\ *)^\top, (*\ 1)^\top \right\}$$
$$2oo2 : \mathcal{VS}_1 = \mathcal{VS}_2 = \mathcal{VS}_3 = \left\{ (1\ 1)^\top \right\}$$
$$2oo3 : \mathcal{VS}_1 = \mathcal{VS}_2 = \mathcal{VS}_3 = \mathcal{VS}_4 = \left\{ (1\ 1\ *)^\top, (*\ 1\ 1)^\top, (1\ *\ 1)^\top \right\}$$

For all four system parts several calculations have been executed with the ASML approach as well as with the IEC 61508 calculation formulas. The results are outlined in table 8.1. The ASML results have been generated by assuming a five year mission time for the respective SIS and conducting proof tests each year for all components simultaneously. As the resulting PFD is cyclic, the averaged value does not depend on the mission time as long as it is chosen as a multiple of the proof test interval. The table entries, i.e., the SIS

Table 8.1: Comparison: ASML approach vs. IEC 61508 formulas ($\beta = 0$, proof test interval 1 year, repair time 8 hours)

Architecture	λ_{du}[FIT]	λ_{dd}[FIT]	$PFD_{avg,\text{IEC 61508}}$	$PFD_{avg,\text{ASML}}$	Difference [%]
1oo1	100	100	$4.40 \cdot 10^{-4}$	$4.39 \cdot 10^{-4}$	−0.14
1oo2	100	100	$2.58 \cdot 10^{-7}$	$2.57 \cdot 10^{-7}$	−0.44
2oo2	100	100	$8.79 \cdot 10^{-4}$	$8.78 \cdot 10^{-4}$	−0.14
2oo3	100	100	$7.74 \cdot 10^{-7}$	$7.69 \cdot 10^{-7}$	−0.70
1oo1	10000	10000	$4.40 \cdot 10^{-2}$	$4.27 \cdot 10^{-2}$	−2.87
1oo2	10000	10000	$2.58 \cdot 10^{-3}$	$2.40 \cdot 10^{-3}$	−7.02
2oo2	10000	10000	$8.79 \cdot 10^{-2}$	$8.29 \cdot 10^{-2}$	−5.71
2oo3	10000	10000	$7.74 \cdot 10^{-3}$	$6.91 \cdot 10^{-3}$	−10.77
1oo1	10000	100	$4.39 \cdot 10^{-2}$	$4.26 \cdot 10^{-2}$	−2.92
1oo2	10000	100	$2.57 \cdot 10^{-3}$	$2.40 \cdot 10^{-3}$	−6.60
2oo2	10000	100	$8.78 \cdot 10^{-2}$	$8.28 \cdot 10^{-2}$	−5.65
2oo3	10000	100	$7.71 \cdot 10^{-3}$	$6.89 \cdot 10^{-3}$	−10.63
1oo1	0	10000	$8.00 \cdot 10^{-5}$	$8.00 \cdot 10^{-5}$	0.00
1oo2	0	10000	$1.28 \cdot 10^{-8}$	$6.40 \cdot 10^{-9}$	−50.00
2oo2	0	10000	$1.60 \cdot 10^{-4}$	$1.60 \cdot 10^{-4}$	0.00
2oo3	0	10000	$3.84 \cdot 10^{-8}$	$1.92 \cdot 10^{-8}$	−50.00

configurations are chosen in a way that they cover very large and very small parameter values in various combinations. In all cases the ASML results are lower than the IEC 61508 averaged PFDs. This originates from the conservative approximations applied during the derivation of the standard's formulas. The standards assume the TTR for the relevant RFR-DRI cycle to be of type A with the related significant simplifications (e.g., the MacLaurin expansion) as derived in subsec. 3.5.2 (p. 36). Additionally, IEC 61508 systems with multiple channels are calculated based on the averaged unavailability of a 1oo1 system. From these derivations stems a failure in the 1oo2 equation, impacting on the result the more λ_{dd} dominates the overall averaged PFD. This effect can clearly be observed by considering the last four lines of table 8.1. In case of $\lambda_{du} = 0$, the ASML value and the IEC 61508 result differ exactly by a factor two (also mentioned in, e.g., [Hil07b]). Apart from this effect, the calculation results differ only slightly. Further validation could be conducted by utilizing the results from [OA10], who extended the IEC 61508 formulas to arbitrary M and N, in order to generate more test scenarios.

8.4 Extended source validation

A useful extension to the IEC 61508 formulas can be found in, e.g., [Hil07a]. The author derives suitable calculation formulas for heterogeneously instrumented systems. The 1oo2 solution can be parameterized with failure rates for *du* as well as *dd* failures, proof test interval and repair time (see table 8.2 for comparison results). The 2oo3 formula is provided as a simplified version such that only *du* failure rates and the proof test interval can be specified. *dd* failures are neglected. Table 8.2 provides a comparison of ASML

Table 8.2: Comparison: ASML approach vs. extended IEC 61508 formula for 2oo3 system part ($\beta = 0$, proof test interval 1 year)

$\lambda_{du,1}$[FIT]	$\lambda_{du,2}$[FIT]	$\lambda_{du,3}$[FIT]	$PFD_{avg,\text{ext. std.}}$	$PFD_{avg,\text{ASML}}$	Difference [%]
10000	100	100	$5.14 \cdot 10^{-5}$	$4.98 \cdot 10^{-5}$	−3.14
10000	10000	100	$2.61 \cdot 10^{-3}$	$2.44 \cdot 10^{-3}$	−6.48
0	100	1000	$2.56 \cdot 10^{-6}$	$2.55 \cdot 10^{-6}$	−0.31
0	1000	1000	$2.56 \cdot 10^{-5}$	$2.54 \cdot 10^{-5}$	−0.70
0	0	10000	0.00	0.00	0.00

and extended standard formula approach for a 2oo3 system part. All required ASMLGs are set up according to sec. 8.3. Again, only slight differences between respective results can be observed. Remarkable result is the ASML approach delivering the correct result $PFD_{avg} = 0$ for $\lambda_{du,1} = \lambda_{du,2} = 0\,\text{FIT}$. Table 8.3 finally also confirms the observations

Table 8.3: Comparison: ASML approach vs. extended IEC 61508 formula for 1oo2 system part ($\beta = 0$, repair time 8 hours, proof test interval 1 year)

$\lambda_{du,1}$[FIT]	$\lambda_{du,2}$[FIT]	$\lambda_{dd,1}$[FIT]	$\lambda_{dd,2}$[FIT]	$PFD_{avg,\text{ext. std.}}$	$PFD_{avg,\text{ASML}}$	Difference [%]
10000	100	100	100	$2.58 \cdot 10^{-5}$	$2.48 \cdot 10^{-5}$	−3.71
100	100	10000	100	$3.16 \cdot 10^{-7}$	$2.91 \cdot 10^{-7}$	−7.97
10000	100	10000	100	$2.58 \cdot 10^{-5}$	$2.48 \cdot 10^{-5}$	−3.93
0	1000	100	100	$5.85 \cdot 10^{-9}$	$3.50 \cdot 10^{-9}$	−40.21

made so far. All ASML generated results are lower than the respective reference values. The major difference in the last line derives from the same problems like discussed in sec. 8.3.

8.5 Extended ASML features

8.5.1 Maintenance group effect

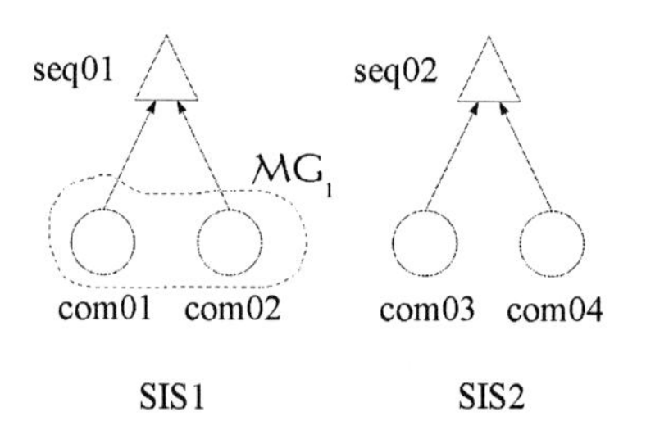

Figure 8.16: DASMLGs of SIS_1 and SIS_2 in the maintenance group example

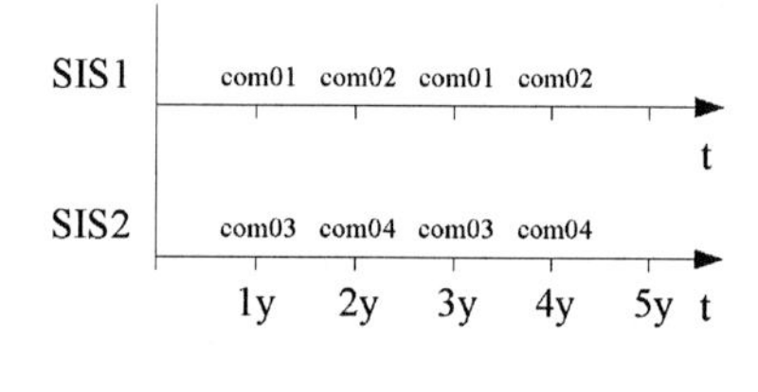

Figure 8.17: SIS_1's and SIS_2's maintenance schemes

The SISs with ASMLGs denoted in fig. 8.16 shall clarify the effect of maintenance groups on a system's unavailability. The SIS marked as SIS_1 has both components in a maintenance group $\mathcal{MG}_1 = \{\text{com01}, \text{com02}\}$. After a successful repair procedure for any of the two components, a proof test is indirectly conducted for the respective other one. SIS_2 does not specify any maintenance group. Figure 8.17 shows the maintenance plan for the SISs, i.e., the points in time a proof test is conducted. Additionally, the corresponding proof tested component is denoted. Components com01 and com03 are proof tested with $\mathcal{PT}_{\text{com01}} = \mathcal{PT}_{\text{com03}} = \{1\,\text{y}, 3\,\text{y}\}$, while com02 and com04 receive maintenance at $\mathcal{PT}_{\text{com02}} = \mathcal{PT}_{\text{com04}} = \{3\,\text{y}, 4\,\text{y}\}$. All components have identical failure rates $\lambda_{du,i} = 100,000\,\text{FIT}$ and repair rates $\mu_{dr,i} = 0.125\,\text{h}^{-1}$. The mission time for the respective SISs is $\tau_{mt} = 5\,\text{y}$. It is expected that SIS_1 performs slightly better with regard to the PFD. Due to the maintenance group effect, a failure in com02 can already be revealed at $t = 1\,\text{y}$ under the condition that the proof tested com01 contains a failure at this point in time.

When comparing both PFDs over discrete time (fig. 8.18 and fig. 8.19), the expectations are totally met. The related averaged PFDs are $PFD_{avg,SIS_1} = 6.1 \cdot 10^{-1}$ and $PFD_{avg,SIS_2} = 7.5 \cdot 10^{-1}$. The maintenance group effect has significant impact on the structure of the transition matrix. While the TMV for SIS_2 is identical to fig. 8.7 (p. 144), the TMV for SIS_1 is depicted in fig. 8.20. Notice, e.g., the transition $(DR\ DU)^{\mathsf{T}} \rightarrow (OK\ DR)^{\mathsf{T}}$, clearly outlining the maintenance group effect: the repair completion for com01 is the initiating event for the repair process of com02.

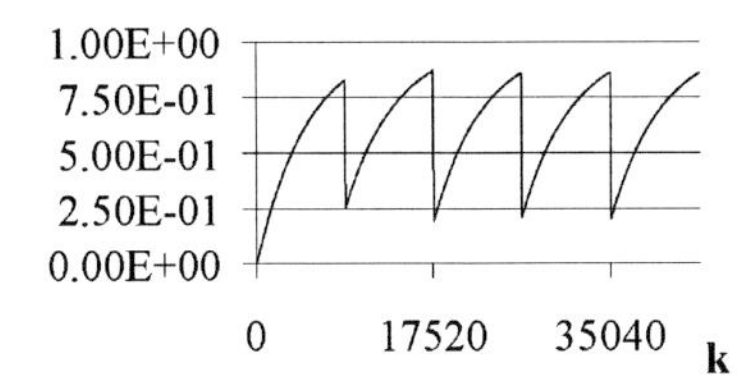

Figure 8.18: PFD over discrete time ($\Delta t = 1\,\mathrm{h}$) for SIS_1

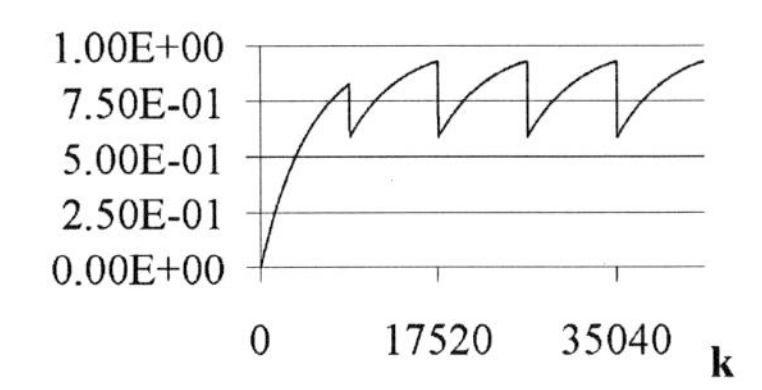

Figure 8.19: PFD over discrete time ($\Delta t = 1\,\mathrm{h}$) for SIS_2

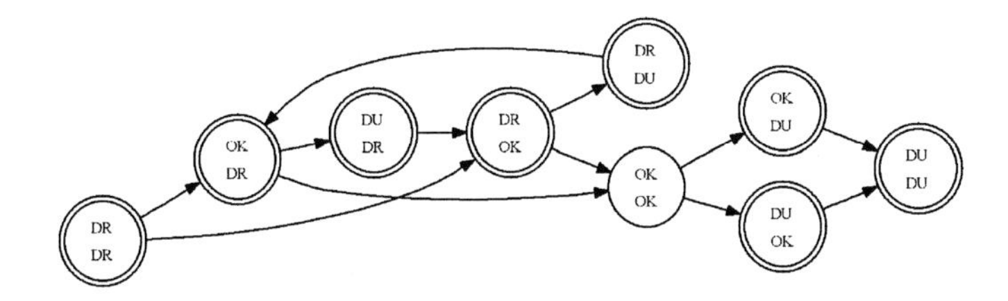

Figure 8.20: TMV for SIS_1

8.5.2 Inhibition effect and *ddx* failures

The ASML approach works on component individual unavailabilities rather than channel unavailability. This allows for the modeling of inhibition effects. Consider two SISs with ASMLGs according to fig. 8.21. SIS_1 is parameterized with $\lambda_{du,\mathrm{com01}} = \lambda_{du,\mathrm{com02}} = \lambda_{du,\mathrm{com03}} = 10,000\,\mathrm{FIT}$, $\lambda_{dd,\mathrm{com01}} = \lambda_{dd,\mathrm{com02}} = \lambda_{dd,\mathrm{com03}} = 10,000\,\mathrm{FIT}$, $\mu_{dr,\mathrm{com01}} = \mu_{dr,\mathrm{com02}} = \mu_{dr,\mathrm{com03}} = 0.125\,\mathrm{h}^{-1}$ and $\mu_{ddr,\mathrm{com01}} = \mu_{ddr,\mathrm{com02}} = \mu_{ddr,\mathrm{com03}} = 0.125\,\mathrm{h}^{-1}$. For SIS_2 the parameters are chosen according to $\lambda_{du,\mathrm{com04}} = \lambda_{du,\mathrm{com05}} = \lambda_{du,\mathrm{com06}} = 10,000\,\mathrm{FIT}$, $\lambda_{ddx,\mathrm{com04}} = \lambda_{ddx,\mathrm{com05}} = \lambda_{ddx,\mathrm{com06}} = 10,000\,\mathrm{FIT}$, $\mu_{dr,\mathrm{com01}} = \mu_{dr,\mathrm{com02}} = \mu_{dr,\mathrm{com03}} = 0.125\,\mathrm{h}^{-1}$ and $\mu_{ddr,\mathrm{com01}} = \mu_{ddr,\mathrm{com02}} = \mu_{ddr,\mathrm{com03}} = 0.125\,\mathrm{h}^{-1}$. On ASML-SIS layer, one maintenance group is specified for each of the two SISs: $\mathcal{MG}_{SIS_1} = \{\mathrm{com01}, \mathrm{com02}, \mathrm{com03}\}$, $\mathcal{MG}_{SIS_2} = \{\mathrm{com04}, \mathrm{com05}, \mathrm{com06}\}$. All components in both SISs are proof tested simultaneously at $\mathcal{PT}_i = \{1\,\mathrm{y}, 2\,\mathrm{y}, 3\,\mathrm{y}, 4\,\mathrm{y}\}$.

Both SISs seem to be identical at first glance. The relevant difference lies in the extended communication for detected failures in SIS_2. As information about a detected failure needs not to be forwarded by subsequently allocated subsystems in the local channel, inhibition effects cannot occur. The relevant process here is as follows: any *du* failure in seq01 or seq02 prevents it from performing its intended safety function. Occurring *ddx* failures for SIS_2 always lead to the initiation of a repair process. Upon completion of the repair, the maintenance group effect (i.e., functional test for the repaired channel) leads to the revelation of all additional hidden *du* failures. In SIS_1, the occurring *dd* failures might get inhibited and thus not reach the ASMLG's output vertex. Hence, the

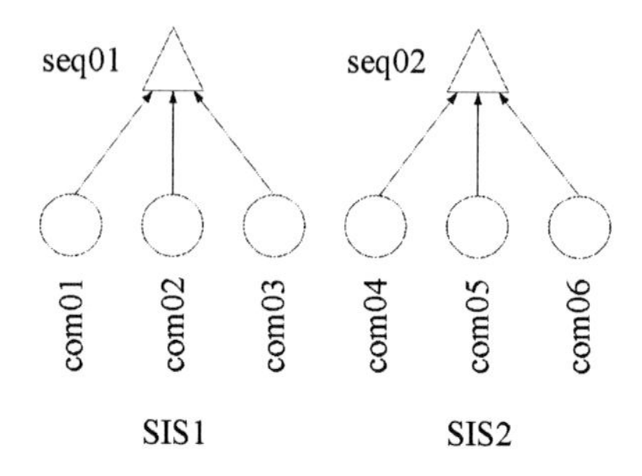

Figure 8.21: ASMLG for SIS_1 and SIS_2 in the inhibition effect example

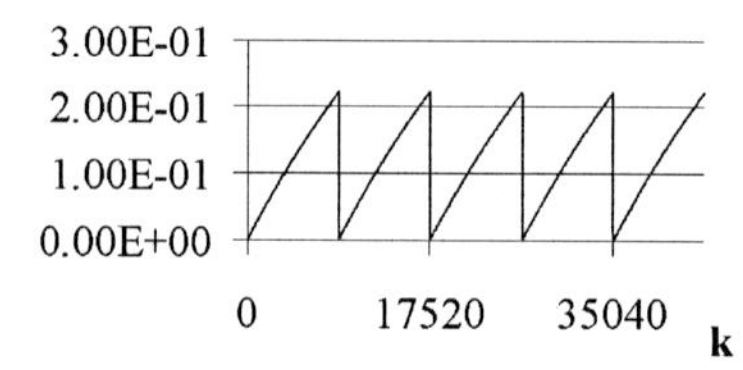

Figure 8.22: PFD over discrete time ($\Delta t = 1\,\mathrm{h}$) for SIS_1 in the *ddx* failure example

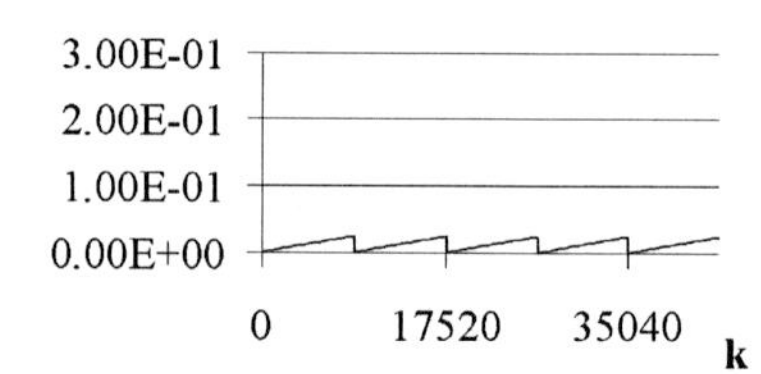

Figure 8.23: PFD over discrete time ($\Delta t = 1\,\mathrm{h}$) for SIS_2 in the *ddx* failure example

potentially contained *du* failures remain in the sequence until the next proof test. It is therefore expected, that SIS_2 performs better than SIS_1. Figures 8.22 and 8.23 depict the related PFDs over discrete time for a mission time of $\tau_{mt} = 5\,\mathrm{y}$. The difference between both systems is significant. The related averaged PFDs are $PFD_{avg,\mathrm{SIS}_1} = 1.18 \cdot 10^{-1}$ and $PFD_{avg,\mathrm{SIS}_2} = 1.26 \cdot 10^{-2}$. Figures 8.24 and 8.25 show excerpts from the TMVs related to the two SISs. Consider the transition $(OK\ OK\ DR)^\mathsf{T} \rightarrow (DDU\ OK\ DR)^\mathsf{T}$ in fig. 8.24. Here, the inhibition effect prevents com01 from entering its DDR state. A second inhibited failure occurs with $(DDU\ OK\ DR)^\mathsf{T} \rightarrow (DDU\ DDU\ DR)^\mathsf{T}$. On the other hand, consider fig. 8.25: the transition $(DR\ OK\ DR)^\mathsf{T} \rightarrow (DR\ DDR\ DR)^\mathsf{T}$ would not be possible in SIS_1. A pure *dd* failure without external communication would result in a transition $(DR\ OK\ DR)^\mathsf{T} \rightarrow (DR\ DDU\ DR)^\mathsf{T}$, as the DR failure in com06 inhibits the *dd* failure at com05.

8.5.3 Partial proof test effect and non-detectable failures

Consider two ASMLSISs, both having their respective ASMLG consist of a single component only. Let com01 in SIS_1 be parameterized according to $\lambda_{du,\mathrm{com01}} = 10,000\,\mathrm{FIT}$,

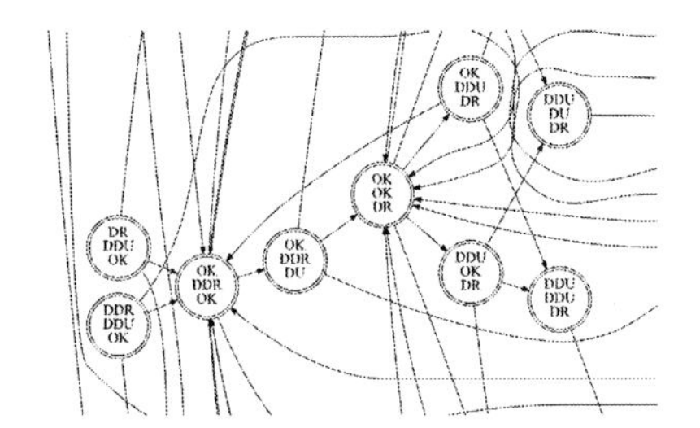

Figure 8.24: Excerpt from the TMV for SIS_1 in the *ddx* failure example

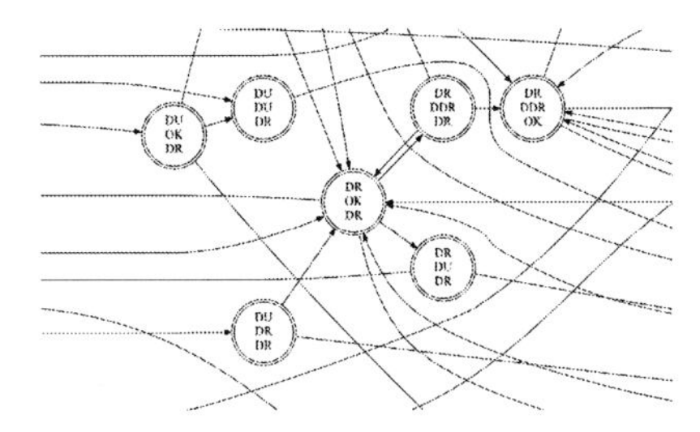

Figure 8.25: Excerpt from the TMV for SIS_2 in the *ddx* failure example

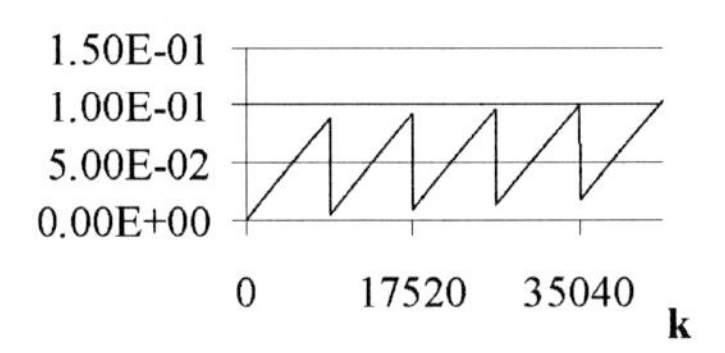

Figure 8.26: PFD over discrete time ($\Delta t = 1\,\text{h}$) for SIS_1 in the example on *dup* and *dn* failures

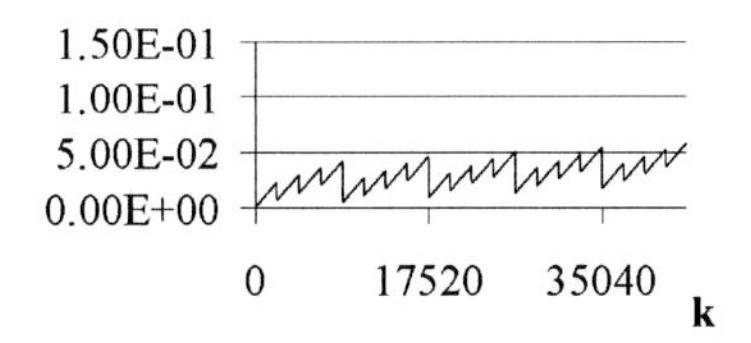

Figure 8.27: PFD over discrete time ($\Delta t = 1\,\text{h}$) for SIS_2 in the example on *dup* and *dn* failures

$\lambda_{dn,\text{com01}} = 500\,\text{FIT}$, $\mu_{dr,\text{com01}} = 0.125\,\text{h}^{-1}$, $PT_{\text{com01}} = \{1\,\text{y}, 2\,\text{y}, 3\,\text{y}, 4\,\text{y}\}$. com02 in SIS_2 has 75 % of its *du* failure rate detectable by a partial proof test. This partial proof test is executed with a frequency four times higher than the standard proof test. The configuration for com02 therefore derives as $\lambda_{du,\text{com02}} = 2,500\,\text{FIT}$, $\lambda_{dup,\text{com02}} = 7,500\,\text{FIT}$, $\lambda_{dn,\text{com02}} = 500\,\text{h}^{-1}$, $\mu_{dr,\text{com02}} = 0.125\,\text{h}^{-1}$, $PT_{\text{com02}} = \{1\,\text{y}, 2\,\text{y}, 3\,\text{y}, 4\,\text{y}\}$,

$$PPT_{\text{com02}} = \left\{ \begin{array}{l} 3\,\text{mths}, 6\,\text{mths}, 9\,\text{mths}, 15\,\text{mths}, 18\,\text{mths}, 21\,\text{mths}, 27\,\text{mths}, 30\,\text{mths}, \\ 33\,\text{mths}, 3\,\text{mths}, 39\,\text{mths}, 42\,\text{mths}, 45\,\text{mths}, 51\,\text{mths}, 54\,\text{mths}, 57\,\text{mths} \end{array} \right\}.$$

Figures 8.26 and 8.27 depict the related PFDs over discrete time. As for SIS_1 and SIS_2 the failure rate for *dn* failures is not equal to zero and no mission time replacements are performed, both curves slightly increase over the whole mission time of five years. After each year, a proof test is conducted for both SISs such that the PFD curve is reset back as far as the non detectable failure fraction allows. The partial proof tests conducted for SIS_2 can be identified as the small decreases each third month. As expected, the averaged PFD for SIS_2 is lower than for SIS_1: $PFD_{avg,\text{SIS}_1} = 5.26 \cdot 10^{-2}$, $PFD_{avg,\text{SIS}_2} = 2.94 \cdot 10^{-2}$.

8.5.4 Degrading effect

Degrading enables for an adaptation of the executed voting of an SPLC while one or more sensor channels contain dangerous detected failures. This provides the possibility to keep the desired amount of redundancy even in case of failures. A typical implementation is a 2oo3 voting scheme that degrades to 1oo2 (and thus a similar safety related availability as 2oo3) and finally performs a shutdown in case of more than one *dd* failure. Since the degrading is only active while the *dd* failures are pending, the time in degradation mode depends on the repair time as well as the failure rate for dangerous detected failures. Predestined for degradation are thus, e.g., process analysis technology (PAT) applications meeting both criteria.

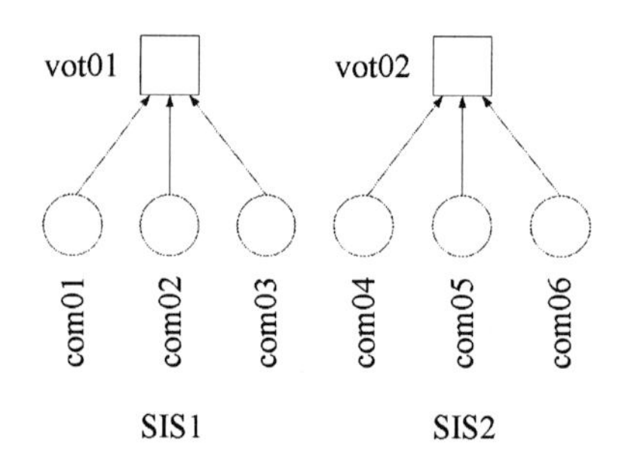

Figure 8.28: ASMLG for two SISs

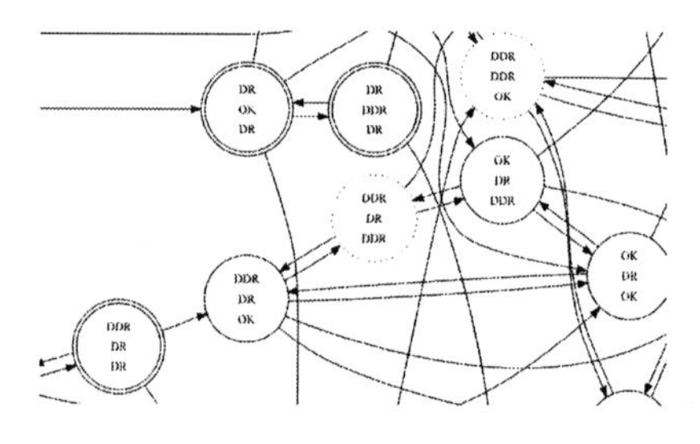

Figure 8.29: Excerpt from the TMV of SIS_2

Consider fig. 8.28 for the related ASMLGs of two SISs SIS_1 and SIS_2. All components com01 to com06 are parameterized equivalently with $\lambda_{dd,i} = \lambda_{du,i} = 10,000\,\text{FIT}$ and $\mu_{dr,i} = \mu_{ddr,i} = 0.00125\,\text{h}^{-1}$, i.e., a very low repair rate resulting in an expected repair time of 800 hours. None of the components receives maintenance over the total mission time of five years for each of the SISs. SIS_1 performs a static 2oo3 voting according to

$$\boldsymbol{vs}_{\text{vot01}} = (\mathcal{VS}_{\text{vot01},1}\ \mathcal{VS}_{\text{vot01},2}\ \mathcal{VS}_{\text{vot01},3}\ \mathcal{VS}_{\text{vot01},4})^\top \text{ with}$$
$$\mathcal{VS}_{\text{vot01},1} = \mathcal{VS}_{\text{vot01},2} = \mathcal{VS}_{\text{vot01},3} = \mathcal{VS}_{\text{vot01},4} = \left\{(1\,1\,*)^\top, (1\,*\,1)^\top, (*\,1\,1)^\top\right\}.$$

SIS_2 degrades the initial 2oo3 voting scheme in case of a detected failure at any input as described above:

$$\boldsymbol{vs}_{\text{vot02}} = (\mathcal{VS}_{\text{vot02},1}\ \mathcal{VS}_{\text{vot02},2}\ \mathcal{VS}_{\text{vot02},3}\ \mathcal{VS}_{\text{vot02},4})^\top \text{ with}$$
$$\mathcal{VS}_{\text{vot02},1} = \left\{(1\,1\,*)^\top, (1\,*\,1)^\top, (*\,1\,1)^\top\right\},$$
$$\mathcal{VS}_{\text{vot02},2} = \left\{(1\,*\,*)^\top, (*\,*\,1)^\top, (*\,1\,*)^\top\right\},$$
$$\mathcal{VS}_{\text{vot02},3} = \mathcal{VS}_{\text{vot02},4} = \left\{(*\,*\,*)^\top\right\}.$$

The calculated averaged PFDs are $PFD_{avg,\text{SIS}_1} = 1.19 \cdot 10^{-1}$ and $PFD_{avg,\text{SIS}_2} = 1.13 \cdot 10^{-1}$ respectively, i.e., a difference of around 5 % in the favor of the degraded system. Figure 8.29 provides the TMV for SIS_2. Safe states (i.e., PFS relevant states) are depicted with a dotted line. The degradation effect can clearly be observed. Consider state $(DDR\, DR\, OK)^\top$. In a static 2oo3 voting, this state space vector represents a state of safety related unavailability, as two components have failed dangerously. But since the single contained DDR state at component com03 forces a degradation, the active voting scheme is $\mathcal{VS}_{\text{SIS}_2,3}$, and since one functional component is left (com06), the considered state is neither PFD nor PFS relevant. A similar consideration can be conducted for $(DDR\, DR\, DDR)^\top$. As two *dd* failures are pending, the voter uses $\mathcal{VS}_{\text{SIS}_2,3}$ as the current voting scheme. This scheme denotes triggering the output of the voter and thus results in a safe state of the SIS.

Chapter 9

Summaries

9.1 Summary in English

Challenges

The aim of this thesis was to provide a suitable method for generically constructing unavailability calculation models for safety loops in process industry. Plant operators have a strong need for numerically calculating safety related and operational unavailability.

The former one is rigidly limited by the international safety standards for the process industry (IEC 61511 and IEC 61508). The average safety related unavailability must be lower than a specific safety integrity level (SIL) that depends on the risk, covered by the considered safety instrumented system (SIS). The proof of SIL-conformity is mandatory since 2003.

On the other hand, operational unavailability leads to unintended SIS trips and thus might cause plant shutdowns. High costs of idleness, expensive cleanup processes, and safety critical startup procedures are the consequences. This characteristic is therefore to be minimized as far as possible. Both safety related and operational unavailability are partially conflictive.

Currently, safety instrumented systems are mathematically treated by referring to pre-derived calculation formulas for SIS parts or by constructing suitable calculation models from scratch, using the recommended mathematical methods suggested by the international safety standards. The former one dramatically limits the degrees of freedom for the engineering process and requires significant simplifications of the SIS structure in order to make the formulas applicable. The latter one quickly leads to large and complicated mathematical structures. Additionally, the standards do not provide information on how to implement a large variety of relevant aspects of modern SISs with the permitted mathematical approaches.

The ASML approach

As a solution to these problems, a generic approach was proposed in this work that automatically constructs suitable calculation models from a formal description. This concept relieves safety engineers from the challenge of manually deriving calculation models. Two positive effects arise from that: firstly, all inherently and explicitly contained complexity of SISs is identified once and condensed into the set of transformation formulas generating the actual calculation models. Secondly, providing a flexible and extensive description language encourages smart engineering processes with non-standard solutions.

A detailed investigation of recent trends in good engineering practice of modern SISs was conducted, serving as a starting point. Relevant effects and parameters were identified. Herefrom, a list of requirements and assumptions was compiled, serving as a basis for the choice of a suitable mathematical method for the approach presented in this work. In the course of the investigation of available methods, multiphase markov models appeared to be most suitable, as they allow for the most detailed representation of the typical repair-failure-repair cycles in process industry: instant repair initiation IRI, e.g., for *dd* failures, and delayed repair initiation DRI, e.g., for *du* failures.

As currently no standardized description language for SISs exists, ASML as a formal description language for safety loops was introduced. This abstract safety markup language allows for describing all identified relevant aspects such as, e.g., inhibition, maintenance groups, partial proof testing, imperfect proof tests, and degrading, in detail, together with all relevant component parameters such as failure and repair rates. Moreover, the representable internal SIS structures, i.e., the interconnections of components (including signal replication), are only slightly restricted. ASML thus provides a high number of degrees of freedom for the engineering process. This clearly contributes to finding a solution to the optimization problem of minimizing a SIS's operational unavailability while keeping the required SIL.

The complete description of a SIS in ASML is called an ASMLSIS. Integral part of it is the ASML graph. The principle of ASMLGs encourages a plain tool-based and graphical engineering process for generating the required description. An ASML description does not require any knowledge about the desired target calculation model type at all.

A set of generator equations were provided that transform an ASMLSIS into an SMPMM, i.e., a safety multiphase markov model. This newly introduced discrete time calculation model type is an extension to the basic multiphase approach and provides additional features in order to replicate the large variety of specific SIS effects as described above. Along with the generator formulas, suitable solver algorithms were presented that allow for retrieving all of the desired unavailability characteristics. These characteristics include the average probability of failure on demand (PFD), i.e., the average safety related unavailability as defined by the standards, but also allow for the calculation of the probability of fail-safe (PFS), i.e., the operational unavailability as economical indicator. The latter one was introduced as the probability of meeting the SIS in a state where it has performed a spurious trip, i.e., where it has unintendedly executed its safety function. As both characteristics are subject to heavy mutual dependencies, it is not possible to calculate them independent from each other. The generated SMPMM therefore holds information on both PFD_{avg} and PFS_{avg}.

Optimization approaches

As markov-based approaches tend to deliver very large models, a set of optimization methods was provided. They render the ASML approach realistically applicable for industrial challenges, as they aim on reducing the markov state space by orders of magnitude.

The most remarkable method here is the decomposition of the original ASMLG, as contained in an ASMLSIS, into several subsystems by algorithmically identifying stochastically independent parts of the SIS. With this technique, an ASMLSIS can be transformed into a decomposed ASMLSIS. A DASMLSIS usually allows for conducting calculations with a markov state space reduced by several orders of magnitude.

Other optimization approaches aim on removing unreachable states or improve computational efficiency by introducing a sparse matrix implementation.

Most of the optimization methods are lossless, i.e., significantly increase calculation speed while reducing memory consumption without impacting on the accuracy of the results.

Validation

By constructing several concrete SISs with predictable calculation outcomes, the functional capability of the ASML approach was demonstrated with regard to all relevant effects. Cross checks with qualified sources for explicit calculation data were conducted. The outlined comparisons demonstrated highly satisfactory performance of the automatically generated SMPMMs and therefore the ASML approach in total.

Outlook

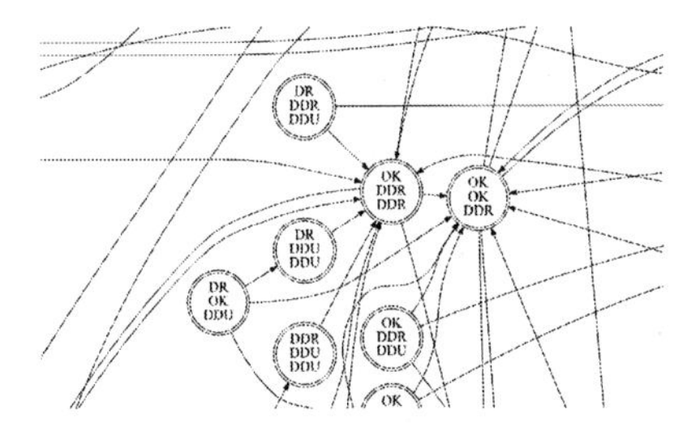

Figure 9.1: TMV for an SMPMM with further optimization potential

Consider fig. 9.1. The depicted TMV is related to SIS_2 in subsec. 8.5.4 (p. 154). Notice vectors with $(* * DDU)^{\top}$. All related markov states are unreachable since the underlying components are directly linked to a voter such that *dd* failures can technically not be inhibited. It makes therefore no sense to include these state space vectors into the state space.

This exemplary observation makes clear that the lossless state space reduction as described in sec. 7.3 (p. 133) can be extended further. Additional criteria in the related equations could identify scenarios where components are immediately linked to voters or the system output - be it a direct link or via a sequence for which the considered component is the last input vertex. The potential in state space reduction is less than a magnitude but is worth the effort - the required algorithms are straightforward and integrate perfectly into the already defined set of equations.

It is to be observed whether the beta model will survive over the next years or whether a more general common cause failure model will be established as standard. If the latter one is the case, then an implementation of that specific approach into the transformation formulas is worthwhile, as workarounds like presented in subsec. 5.5.4 (p. 102) currently result in additional components and thus contribute to the state space explosion.

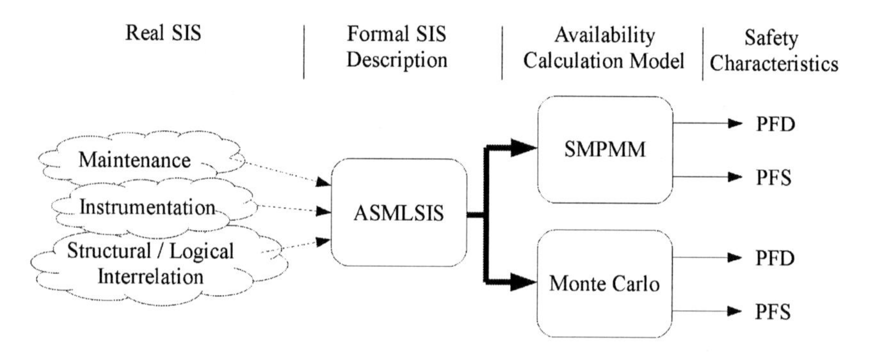

Figure 9.2: Flow diagram of the ASML approach with additional monte carlo transformation

As sec. 4.5 (p. 60) clearly pointed out the positive aspects of monte carlo based approaches, a matching set of transformation formulas is worthwhile a consideration. A good starting point is the CFSM from subsec. 5.2.3 (p. 73). This automaton can easily be transformed into a PN. A major advantage of the MC approach in combination with PNs is the simple implementation of an explicit number of maintenance teams. The drawbacks (confidence intervals instead of sharp unavailabilities etc.) have been identified in sec. 4.6 (p. 66). In fig. 9.2, the extended ASML approach is depicted.

9.2 Kurzfassung in deutscher Sprache (Extended summary in German)

Problemumfeld

Safety Instrumented Systems (SIS) werden international in der Prozessindustrie eingesetzt. Sie haben die Aufgabe, gefährliche Betriebszustände zu detektieren und, im Falle einer Anforderung, einen vordefinierten sicheren Zustand herzustellen. Anlagenbetreiber haben ein besonderes Interesse daran, zwei spezielle Kenngrößen von SIS numerisch bestimmen zu können: Sicherheitstechnische und betriebstechnische Unverfügbarkeit.

Die erste Größe repräsentiert die Wahrscheinlichkeit, das SIS zu einem beliebigen Zeitpunkt in einem Zustand vorzufinden, in dem es nicht mehr in der Lage ist, seine zugedachte Schutzfunktion korrekt auszuführen. Die relevanten internationalen Standards IEC 61511 und IEC 61508 bezeichnen die mittlere sicherheitstechnische Unverfügbarkeit als PFD_{avg} (Average Probability of Failure on Demand). Jede PLT-Schutzeinrichtung (so die bisherige Bezeichnung für SISs) muss - entsprechend dem durch sie abgedeckten Risiko - harte Grenzwerte für PFD_{avg} einhalten. Diese werden in den Sicherheitsnormen als Safety Integrity Levels (SIL) bezeichnet. Sie setzen eine quantitative Bestimmung der sicherheitstechnischen Unverfügbarkeit voraus.

Die zweite relevante Kenngröße, die betriebstechnische Unverfügbarkeit, ist von wirtschaftlichem Interesse. Sie bezieht sich auf die Wahrscheinlichkeit, ein SIS zu einem beliebigen Zeitpunkt in seinem zugedachten sicheren Zustand vorzufinden. Die Ursache hierfür ist keine Anforderung aus dem Prozess, sondern ein aktiver SIS-interner Fehler. Es handelt sich also um eine ungewollte Auslösung der Schutzfunktion. Aufgrund der Verkettung von Schutzeinrichtungen untereinander, kann eine einzelne Fehlauslösung zur Schnellabschaltung der gesamten Anlage führen und somit hohe Stillstandskosten verursachen. Auch durch die Schnellabschaltung verursachte Anlagenschäden bzw. Kosten für unbrauchbar gewordenes Produkt sind ein für Betreiber meist erheblicher finanzieller Aufwand. Jede Abschaltung erfordert einen Wiederanfahrprozess. Aus sicherheitstechnischer Sicht ist dieser als kritisch zu beurteilen (meist manuelle Durchführung, keine Routine), der nach Möglichkeit nur selten durchgeführt werden sollte. Die betriebstechnische Unverfügbarkeit ist somit ein Gradmesser für Kosten durch Fehlauslösung von SIS. Da die Sicherheitsstandards wirtschaftliche Kenngrößen nicht behandeln, wird in dieser Arbeit - in Anlehnung an andere Literaturquellen - die Probability of Fail-Safe (PFS) eingeführt. Auch hier kann eine mittlere Unverfügbarkeit, die PFS_{avg}, angegeben werden.

Während die PFD_{avg} nur unter die durch IEC 61511 und IEC 61508 vorgegebenen Grenzwerte gebracht werden muss, haben Anlagenbetreiber ein verständliches Interesse daran, die PFS_{avg} möglichst zu minimieren.

Relevante Effekte in modernen PLT-Schutzeinrichtungen

In einem Überblick über aktuelle Trends und Methoden beim SIS-Design, werden wichtige Effekte und Aspekte zusammengestellt und analysiert, die für die numerische Berechnung von PFD_{avg} und PFS_{avg} von Bedeutung sind. Die meisten der hier genannten Effekte werden derzeit in der übrigen Literatur überhaupt nicht oder nur rudimentär behandelt.

So ist zunächst die Wechselwirkung von aktiven und passiven Fehlern in SISs zu erwähnen. Eine primäre Rolle spielen hierbei zwei Effekte, die mit 'Inhibition' und 'Maintenance Groups' bezeichnet werden. Inhibition tritt auf, wenn in einem Sensorkanal eine Komponente nahe der sicherheitsgerichteten speicherprogrammierbaren Steuerung (SSPS) passiv ausfällt, z.B. durch einen *du*-Fehler. Tritt nun ein aktiver Fehler weiter vorne im Kanal auf, z.B. am Sensorelement, so kann dieses Fehlabschaltesignal die Steuerung aufgrund des *du*-Fehlers nicht erreichen. Ein sonst selbstmeldender Fehler bleibt somit vorläufig unterdrückt. Der umgekehrte Effekt betrifft die 'Maintenance Groups'. Fällt das Sensorelement passiv aus, so würde dieser Fehler frühestens bei einer Routinekontrolle (in der Prozessindustrie üblicherweise jährlich) auffallen. Wenn jedoch beispielsweise die zugehörige Eingangskarte an der SSPS einen aktiven Fehler verursacht, so wird für diese Komponente ein Reparaturvorgang initiiert. Nach Abschluss der Reparatur wird natürlich die Funktion der neuen Komponente getestet. Dies kann nicht komponentenindividuell erfolgen. Üblicherweise wird der Sensorkanal softwareseitig kurz gebrückt, und eine Funktionstesttaste am Sensorelement betätigt. Erreicht das initiierte Schaltsignal die Steuerung, so kann von Funktionstüchtigkeit ausgegangen werden. Auf diese Weise kann nun der passive Fehler am Sensorelement vorzeitig - also vor der nächsten Wiederholungsprüfung - aufgedeckt und repariert werden. Weitere Interaktionsmöglichkeiten bestehen.

Fortgeschrittenere Konzepte beziehen sich z.B. auf Partial-Stroke-Tests, unvollständige Wiederholungsprüfungen, komponentenindividuelle Prüfpläne oder Sekundärsensoren, die zur Überwachung bestimmter Fehlerzustände der eigentlichen SIS-Komponenten eingesetzt werden. Aber auch dynamisches Voting, also die Anwendung unterschiedlicher SSPS-Votingalgorithmen in Abhängigkeit der Anzahl erkannter Fehler an den Sensoreingängen, spiegelt die Komplexität moderner PLT-Schutzeinrichtungen wider.

Gegenwärtig werden hauptsächlich die von IEC 61508 angebotenen PFD-Rechenformeln zur Berechnung der PFD_{avg} genutzt. Diese Formeln beinhalten starke Konservatismen und gehen von einigen wenigen und starren Systemarchitekturen aus. So können beispielsweise nicht einmal diversitär instrumentierte Schutzkreise berechnet werden, obwohl die Sicherheitsnormen Diversität nachdrücklich empfehlen. Anwender sind dazu gezwungen, eine rechnerisch konservative Anpassung ihrer Systemparametrierung vorzunehmen, um das betrachtete SIS berechenbar zu machen. Hierdurch entsteht die unbefriedigende Situation, dass unter Umständen Redundanzen nachgerüstet werden müssen, da durch die notwendigen Konservatismen der vorgegebene SIL gerade eben überschritten wird. Eine 'korrekte' Berechnung hingegen würde die sicherheitstechnische Eignung der ursprünglichen Schutzeinrichtung nachweisen. Weiterhin legitimieren die Standards verschiedene mathematische Rechenverfahren zur Bestimmung der sicherheitstechnischen Unverfügbarkeit (z.B. Fehlerbäume und Markovmodelle). Für den Betreiber sind diese Methoden wenig brauchbar, da unklar ist, wie die relevanten Effekte umzusetzen sind. Es ist nicht einmal klar, ob überhaupt alle Verfahren geeignet sind, die Effekte nachzubilden. Nahezu alle Ansätze erfordern mathematisch geschultes Fachpersonal, das in kleineren Betrieben üblicherweise nicht zur Verfügung steht. Zusätzlich haben die vielversprechendsten Verfahren die Eigenschaft, sehr schnell sehr große und unübersichtliche Modelle zu produzieren, so dass eine händische Erstellung bei gleichzeitiger Beherrschung der Komplexität nicht möglich ist. Sobald drei und mehr ausfallfähige Objekte (Komponenten oder Kanäle) betrachtet werden müssen, scheitert dieser Ansatz. So sind einerseits die Möglichkeiten, andererseits die Freiheitsgrade, beim SIS-Design derzeit stark eingeschränkt.

ASML-Ansatz

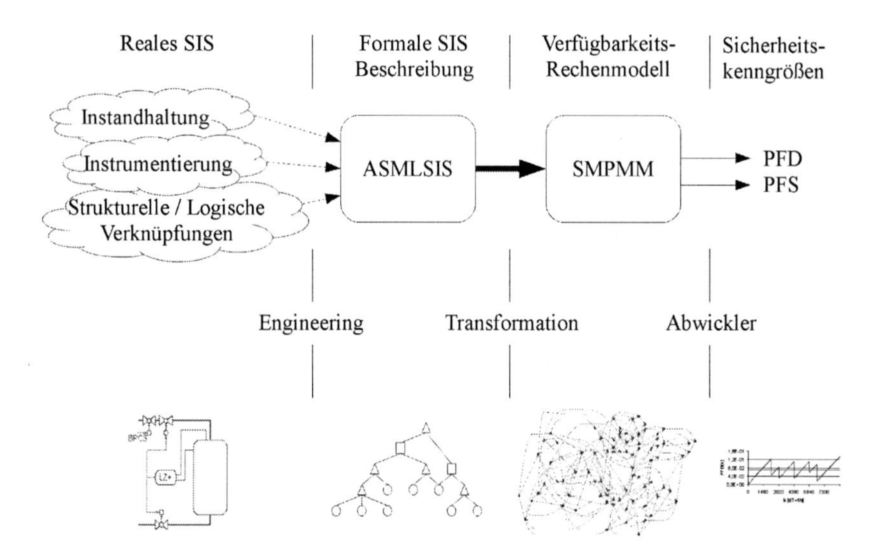

Abbildung 9.3: Der ASML-Ansatz im Überblick

Der in dieser Arbeit vorgestellte ASML-Ansatz (ASML steht für Abstract Safety Markup Language) stellt einen ganzheitlichen Lösungsansatz für das beschriebene Problemfeld dar (siehe Abb. 9.3). Er besteht im Wesentlichen aus drei Teilen:

- Beschreibung einer PLT-Schutzeinrichtung unter Berücksichtigung sämtlicher relevanter Effekte mit der formalen Beschreibungssprache ASML. Ein so spezifizierter Schutzkreis heißt ASMLSIS. Wichtigster Bestandteil eines ASMLSIS ist der ASML Graph (ASMLG, siehe Abb. 9.4).
- Automatische Generierung eines geeigneten Rechenmodells aus einem ASMLSIS. Im vorgestellten Ansatz sind dies Safety Multiphase Markov Models (SMPMMs).
- Ableitung der Kenngrößen PFD_{avg} und PFS_{avg} aus dem Rechenmodell.

Durch diese Vorgehensweise kann zunächst der Bedarf an mathematisch geschultem Fachpersonal egalisiert werden. Der benötigte Engineeringaufwand reduziert sich auf die Erstellung der formalen Beschreibung mit ASML. Diese Beschreibung erfolgt zum größten

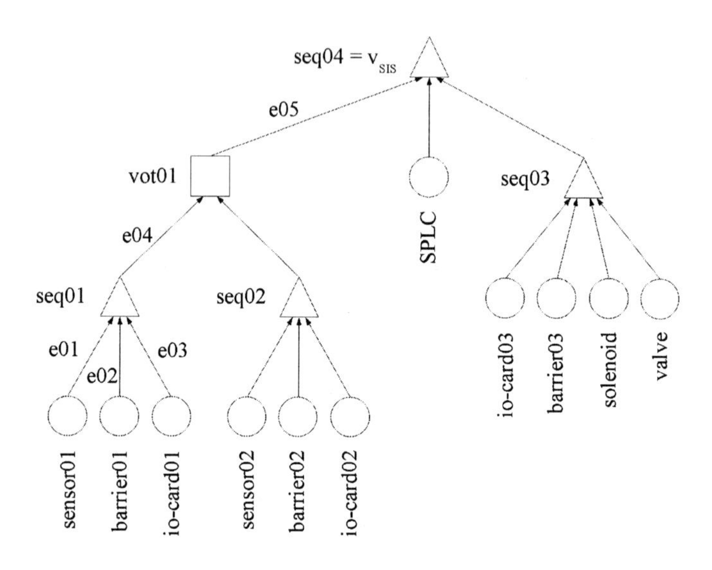

Abbildung 9.4: Beispiel-ASMLG zu einem SIS mit zweikanaliger Sensorik und einkanaligem Aktorteilsystem

Teil grafisch über den ASMLG und erfordert keinerlei detaillierteren Kenntnisse über beispielsweise stochastische Abhängigkeiten oder die mathematische Darstellung oben genannter Effekte. ASML ist somit eine reine Beschreibungssprache. Im Anschluss wird automatisch, mit Hilfe geeignet entworfener Generatorformeln, ein Rechenmodell erzeugt, aus dem anschließend die beiden relevanten Unverfügbarkeitskenngrößen bestimmt werden können. Erwähnenswert hierbei ist, dass die Generatorformeln mit im Prinzip beliebig strukturierten ASMLSIS beliebiger Größe zurechtkommen. Solange eine Schutzfunktion in ASML beschreibbar ist, kann auch das zugehörige Rechenmodell erzeugt werden.

Zur Umsetzung dieses Lösungskonzepts werden in der Arbeit zunächst die gängigsten Rechenmodelle vorgestellt, miteinander verglichen, und anschließend im Hinblick auf ihre Brauchbarkeit für den ASML-Ansatz bewertet. Hierbei stellt sich heraus, dass nur Markovmodelle den aufgestellten Anforderungen gerecht werden können. Bedingt durch die spezielle Eigenschaft der Prozessindustrie, periodische Funktionsprüfungen durchzuführen, eignet sich eine spezielle Unterart von Markovmodellen besonders für das zu realisierende Konzept: diskrete Multiphasen-Markovmodelle. Diese sind in der Lage, die Information, dass die sicherheitstechnische Unverfügbarkeit einer Komponente unmittelbar nach einer Prüfung null ist (unter Annahme perfekter Wiederholungsprüfungen), unmittelbar abzubilden. Im Gegensatz zu anderen Modellvarianten ist hier also auch die Punkt-Unverfügbarkeit, also beispielsweise die PFD(t), sinnvoll bestimmbar.

Um den speziellen Anforderungen des ASML-Konzeptes gerecht zu werden, wird ein neuer Typ diskretes Multiphasen-Markovmodell eingeführt, das SMPMM (Safety Multiphase Markov Model):

$$\mathcal{SMPMM}\left(\mathcal{S}, \boldsymbol{A}, \boldsymbol{p}_0, \mathcal{T}, k_{mt}, \boldsymbol{c}_{PFD}^{\top}, \boldsymbol{c}_{PFS}^{\top}\right).$$

Dieses enthält sämtliche Informationen um PFD_{avg} und PFS_{avg} eines SISs bestimmen zu können. Aufgrund der wechselseitigen Abhängigkeiten der Kenngrößen ist eine getrennte Berechnung, z.B. über zwei Modelle, nicht möglich.

Für die Transformation eines ASMLSIS in ein SMPMM werden in dieser Arbeit geeignete Formeln bereitgestellt. Mit ihnen können der Markov-Zustandsraum aufgespannt ($\mathcal{S}$), Transitionsmatrix ($\boldsymbol{A}$) und Verschiebungsmatrizen ($\mathcal{T}$) bestimmt werden. Eine wesentliche Herausforderung ist die Bestimmung der Selektionsvektoren ($\boldsymbol{c}_{PFD}^{\top}$ und $\boldsymbol{c}_{PFS}^{\top}$). Diese wählen diejenigen Markovzustände aus, deren Aufenthaltswahrscheinlichkeit beispielsweise zur PFD_{avg} beiträgt. Hierfür ist es notwendig herauszufinden, ob ein bestimmter Markovzustand ein sicherheitstechnisch funktionsfähiges SIS repräsentiert. Dies kann durch Einbezug des zugrundeliegenden ASMLG bewerkstelligt werden. Dabei wird ein kombinatorisch generierter Zustandsvektor des SMPMMs (z.B. $(OK\ DU\ DU)^{\top}$) als eine Ist-Belegung der Komponentenzustände in ASMLGs interpretiert (Komponenten sind in ASMLGs durch Kreise repräsentiert, siehe Abb. 9.4). Durch rekursive Evaluierung des Graphen kann der Systemzustand bei derart vorgegebener Komponentenzustandskonfiguration ermittelt werden. Das Resultat - beispielsweise 'sicherheitstechnisch unverfügbar' - kann unmittelbar zur Bewertung des zugehörigen Markovzustands verwendet werden.

Anschließend können die Kenngrößen PFD_{avg} und PFS_{avg} aus einem SMPMM analog zu einem gewöhnlichen Multiphasen-Markovmodell abgeleitet werden. Für die Mittelwertbildung wird eine SIS-Einsatzdauer k_{mt} aus dem zugehörigen ASMLSIS übernommen.

Implementierungstechnische Optimierung

Markovmodell-basierte Rechenmodelle leiden unter exponentiellem Wachstum des Zustandraums für linear wachsende Anzahl simulierter Komponenten. Um den vorgestellten ASML-Ansatz für die Praxis zu ertüchtigen, werden mehrere Optimierungsansätze vorgestellt. Diese heben einerseits darauf ab, die Zustandsraumexplosion um Größenordnungen zu dämpfen, und andererseits die Modell-Abwicklung, d.h. Berechnungszeit und Speicherverbrauch, zu minimieren. Dabei sind die meisten Optimierungsverfahren gleichzeitig verlustfrei, d.h. sie wirken sich nicht auf das Rechenergebnis im Sinne von Ungenauigkeit (z.B. approximativ) aus.

Mit der ASML-Dekomposition können in einem ASMLSIS stochastisch unabhängige Teilsysteme automatisch detektiert und extrahiert werden. Die Unverfügbarkeit solcher Subsysteme kann über getrennte SMPMMs berechnet werden. Über geeignet bereitgestellte Kombinatorikformeln werden die einzelnen Subsystem-Unverfügbarkeiten anschließend zur SIS-Unverfügbarkeit verknüpft. Ein solches, durch Dekomposition entstehendes System, wird als Decomposed ASMLSIS (DASMLSIS) bezeichnet. Die Anwendbarkeit in der Praxis wird positiv bewertend diskutiert und reduziert die Zustandsraumgröße und damit auch den Rechenaufwand signifikant.

Mit dem Prinzip der verlustfreien Zustandsraumreduktion können unerreichbare Markovzustände identifiziert und aus dem Rechenmodell entfernt werden. Auch hierbei werden Zustandsraum und damit Rechenzeit, sowie Speicherverbrauch verringert.

Die Transitionsmatrizen der SMPMM der betrachteten technischen Systeme sind extrem spärlich besetzt (typischerweise weniger als 1 %). Durch geeignete Implementierung dieser Matrizen können die laufzeitbestimmenden Matrix-Multiplikationen um mehrere Größenordnungen gegenüber der Standardimplementierung beschleunigt werden.

Die Auswirkung dieser und weiterer Optimierungsansätze wird detailliert untersucht und bewertet.

Validierung der Ergebnisse

Abschließend wird die Implementierung sämtlicher als relevant erachteten Effekte exemplarisch anhand mehrerer Validierungsbeispiele plausibilisiert. Hierzu werden geeignete SIS konstruiert, über deren Unverfügbarkeit a-priori Information verfügbar sind. Diese werden anschließend in ASML modelliert. Die Ergebnisse nach Transformation auf SMPMM und Ableitung der Kenngrößen können mit den Erwartungen verglichen werden. Es zeigt sich hervorragende Übereinstimmung.

Rechenergebnisse aus verfügbaren autorisierten Quellen (IEC 61508 und Erweiterungen) werden zum Vergleich herangezogen. Hierzu werden die den Quellen zugrundeliegenden SIS in ASML beschrieben und die hieraus automatisch ermittelten Unverfügbarkeiten den autorisierten Ergebnissen gegenübergestellt. Die durchweg sehr kleine Abweichung der Werte wird ergänzend und erläuternd diskutiert.

Kapitel 10

Appendix

10.1 Type A RFR-DRI process approximation

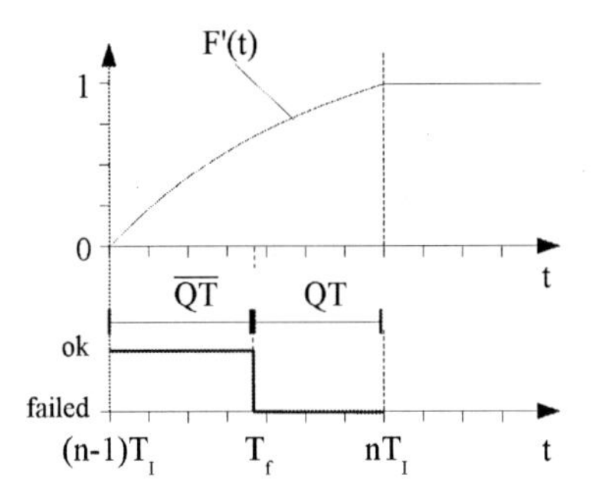

Abbildung 10.1: Derivation of the FR process

It is to be shown that

$$\mathrm{E}(QT) = T_I - \frac{1}{\lambda} + \frac{T_I}{\exp(\lambda T_I) - 1}. \qquad (10.1)$$

A component with exponentially distributed lifetime $\mathrm{F}(t) = 1 - \exp(-\lambda t)$ is periodically tested each nT_I. In case of a failure it is repaired instantly. The component's lifetime thus ends after a time $\overline{QT}$ inbetween two subsequent proof tests. This time is a random variable with PDF $f'_{\overline{QT}}(t)$ and matching CDF $F'_{\overline{QT}}(t)$ (see fig. 10.1). According to [Sch76] and [Sch80], all proof test periods contribute to this random variable (i.e., sum of disjoint failure event probabilities), the PDF can be set up according to

$$f'_{\overline{QT}}(t) = \sum_{n=0}^{\infty} \mathrm{f}(nT_I + t) = \sum_{n=0}^{\infty} \lambda \exp(-\lambda (nT_I + t)).$$

for $0 \leq t < T_I$. $f'_{\overline{QT}}$ is zero outside this interval. The PDF can be expanded into

$$\begin{aligned} f'_{\overline{QT}}(t) &= \lambda \exp(-\lambda t) + \lambda \exp(-\lambda (T_I + t)) + \lambda \exp(-\lambda (2T_I + t)) + \cdots . \\ &= \lambda \exp(-\lambda t) \left[1 + \exp(-\lambda T_I) + \exp(-2\lambda T_I) + \cdots\right]. \end{aligned} \tag{10.2}$$

As generally

$$\sum_{n=0}^{\infty} z^n = \frac{1}{1-z}$$

holds for $|z| < 1$, eq. 10.2 can be written as

$$f'_{\overline{QT}}(t) = \frac{\lambda \exp(-\lambda t)}{1 - \exp(-\lambda T_I)}.$$

The expected queuing time of the component, i.e., its downtime until the conduction of the upcoming proof test, can now be expressed as the expected value of the random variable $QT = T_I - \overline{QT}$ according to:

$$\begin{aligned} \mathrm{E}(QT) &= T_I - \mathrm{E}(\overline{QT}) \\ &= T_I - \int_0^{T_I} t \cdot \frac{\lambda \exp(-\lambda t)}{1 - \exp(-\lambda T_I)} dt. \end{aligned}$$

The result is the proposed eq. 10.1.

10.2 MacLaurin extension

The exponential term $\exp(-\lambda T)$ can be expanded into a Taylor series centered around zero (then called a MacLaurin series) according to

$$\exp(-\lambda T) = \sum_{n=0}^{\infty} \frac{\left[(\exp(-\lambda T))^{(n)}\right]_{T=0}}{n!} (T-0)^n = \sum_{n=0}^{\infty} \frac{(-\lambda)^n}{n!} T^n.$$

The series converges against the exponential term for arbitrary T as the convergence radius R can be calculated from the series' coefficients $\frac{(-\lambda)^n}{n!}$:

$$R = \lim_{n\to\infty} \frac{\left|\frac{(-\lambda)^n}{n!}\right|}{\left|\frac{(-\lambda)^{n+1}}{(n+1)!}\right|} = \lim_{n\to\infty} \frac{(n+1)!}{n!} \frac{\lambda^n}{\lambda^{n+1}} = \lim_{n\to\infty} \frac{(n+1)}{\lambda} = \infty.$$

The first three summands from the MacLaurin series are often used throughout this thesis as approximation for $\exp(-\lambda T)$. The related expansion is

$$\exp(-\lambda T) \approx 1 - \lambda T + \frac{\lambda^2 T^2}{2}.$$

10.3 Superimposed exponential lifetime processes

It is to be shown that the time to failure in a concurrency situation of R exponentially distributed failure processes competing for bringing a component to its failed state, i.e.,

$$TTF' = \min(TTF_1, TTF_2, \ldots, TTF_R),$$

is exponentially distributed with

$$\mathrm{P}(TTF' \leq t) = \mathrm{F}'(t) = 1 - exp\left(-\sum_R \lambda_i t\right).$$

Proof immediately follows from [HS99] as well as [BB95] and is slightly extended to suit the previous phrasing.

$$\begin{aligned}
TTF' &= \min(TTF_1, TTF_2, \ldots, TTF_R) \\
\mathrm{P}(TTF' > t) &= \mathrm{P}(\min(TTF_1, TTF_2, \ldots, TTF_R) > t) \\
&= \mathrm{P}((TTF_1 > t) \cap (TTF_2 > t) \cap \ldots \cap (TTF_R > t)) \\
&= \mathrm{P}(TTF_1 > t) \cdot \mathrm{P}(TTF_2 > t) \cdot \ldots \cdot \mathrm{P}(TTF_R > t)) \\
&= exp(-\lambda_1 t) \cdot exp(-\lambda_2 t) \cdot \ldots \cdot exp(-\lambda_R t) \\
&= exp(-(\lambda_1 + \lambda_2 + \ldots + \lambda_R)t) \\
&= exp(-\sum_R \lambda_i t)
\end{aligned}$$

From $\mathrm{P}(TTF' > t) = exp(-\sum_R \lambda_i t)$, immediately $\mathrm{P}(TTF' \leq t) = 1 - exp(-\sum_R \lambda_i t)$ follows.

Nomenclature

Acronyms

Acronym	Explanation	Reference
a	active	def. 5.2 (p. 73)
AR	active under repair	def. 5.1 (p. 72)
ASML	abstract safety markup language	sec. 1.2 (p. 3)
AU	active unrevealed	def. 5.1 (p. 72)
BPCS	basic process control system	subsec. 2.2.1 (p. 10)
CC	common cause	subsec. 2.3.6 (p. 21)
CDF	cumulative density function	
CPU	central processing unit	
DCS	decentralized control system	
dd	dangerous detected	def. 5.2 (p. 73)
DDR	dangerous detected under repair	def. 5.1 (p. 72)
DDU	dangerous detected unrevealed	def. 5.1 (p. 72)
ddx	dangerous detected with external communicati- on	def. 5.2 (p. 73)
DN	dangerous non-detectable	def. 5.1 (p. 72)
dn	dangerous non-detectable	def. 5.2 (p. 73)
DR	dangerous under repair	def. 5.1 (p. 72)
DT	down time	sec. 3.5 (p. 34)
DU	dangerous undetected	def. 5.1 (p. 72)
du	dangerous undetected	def. 5.2 (p. 73)
DUP	dangerous undetected partial proof test detec- table	def. 5.1 (p. 72)
dup	dangerous undetected partial proof test detec- table	def. 5.2 (p. 73)
EDPC	electronic data processing center	
FIT	failures in time i.e.per 10^9 hours	
FLOP	floating point operation	
FMEA	failure modes and effects analysis	
FR process	failure-repair process	sec. 3.3 (p. 32)
GSPN	generalized stochastic petri nets	subsec. 4.5.3 (p. 62)

Acronym	Explanation	Reference
HAZOP	hazard and operability investigation	sec. 3.1 (p. 29)
IEC	international electrotechnical commission	
LOPA	layers of protection analysis	sec. 3.1 (p. 29)
LS	logic solver = SPLC	subsec. 2.2.2 (p. 14)
MC	monte carlo	sec. 4.5 (p. 60)
MCS	monte carlo simulation	
MDT	mean down time	sec. 3.5 (p. 34)
MT	mission time (used e.g.as subindex)	
MUT	mean up time	sec. 3.5 (p. 34)
MooN	m-out-of-n	subsec. 2.2.2 (p. 14)
NAMUR	Normenarbeitsgemeinschaft für Mess- und Regeltechnik in der Chemischen Industrie	
ODE	ordinary differential equation	
P& ID	piping & instrumentation diagram	
PAT	process analysis technology	subsec. 2.2.1 (p. 10)
PDF	probability density function	
PFD	probability of failure on demand	sec. 3.1 (p. 29)
PFH	probability of failure per hour	
PFS	probability of fail-safe	def. 3.1 (p. 32)
PN	petri net	subsec. 4.5.3 (p. 62)
PRNG	pseudo random number generator	sec. 4.5 (p. 60)
PT	proof test (used e.g.as subindex)	
PPT	partial proof test (used e.g.as subindex)	
PTC	proof test coverage	sec. 2.2.3 (p. 17)
RF process	repair-failure process	sec. 3.3 (p. 32)
RFR cycle	repair-failure-repair cycle	sec. 3.5 (p. 34)
RFR-DRI-A	type A RFR cycle with delayed repair initiation	subsec. 3.5.2 (p. 36)
RFR-DRI-B	type B RFR cycle with delayed repair initiation	subsec. 3.5.3 (p. 38)
RFR-DRI-C	type C RFR cycle with delayed repair initiation	subsec. 3.5.4 (p. 40)
RFR-IRI	RFR cycle with instant repair initiation	subsec. 3.5.1 (p. 36)
SDC	shutdown combination	def. 5.16 (p. 87)
SIF	safety instrumented function	sec. 1.1 (p. 1)
SIL	safety integrity level	sec. 3.1 (p. 29)
SIS	safety instrumented system	sec. 1.1 (p. 1)
SMI	sparse matrix implementation	sec. 7.4 (p. 134)
SPLC	safety programmable logic controller = LS	sec. 1.1 (p. 1)
TMV	transition matrix visualization	chap. 8 (p. 141)
TTF	time to failure	sec. 3.3 (p. 32)
TTL	transistor-transistor logic	
TTR	time to restoration	sec. 3.4 (p. 33)
UT	up time	sec. 3.5 (p. 34)

Objects

Object	Explanation	Reference
$\mathcal{ASMLG}$	abstract safety markup language graph	def. 5.3 (p. 78)
$\mathcal{ASMLSIS}$	abstract safety markup language safety instrumented system	def. 5.22 (p. 98)
$\mathcal{CFSM}$	component finite state machine	subsec. 5.2.3 (p. 73)
$\mathcal{COMBINAT.-SEQUENCE}$	DASMLG combinatorial sequence vertex type	def. 7.4 (p. 125)
$\mathcal{COMPONENT}$	ASMLG component vertex type	def. 5.11 (p. 83)
$\mathcal{DASMLG}$	decomposed abstract safety markup language graph	def. 7.1 (p. 123)
$\mathcal{DASMLSIS}$	decomposed abstract safety markup language safety instrumented system	def. 7.8 (p. 127)
$\mathcal{FT}$	fault tree	subsec. 4.3.1 (p. 45)
$\mathcal{MM}$	markov model	subsec. 4.4.1 (p. 50)
$\mathcal{MPMM}$	multiphase markov model	def. 4.1 (p. 56)
$\mathcal{RBD}$	reliability block diagram	subsec. 4.2.1 (p. 42)
$\mathcal{SEQUENCE}$	ASMLG sequence vertex type	def. 5.13 (p. 85)
$\mathcal{SMPMM}$	safety multiphase markov model	def. 6.1 (p. 106)
$\mathcal{SUBSYSTEM}$	DASMLG subsystem vertex type	def. 7.3 (p. 124)
$\mathcal{TRANS}$	transition in an IEC 61508 ed. 2 petri net	subsec. 4.5.3 (p. 62)
$\mathcal{VOTER}$	ASMLG voter vertex type	def. 5.14 (p. 85)

Variables

Variable	Explanation	Reference
$\boldsymbol{A}$	transition matrix of a markov model	subsec. 4.4.1 (p. 50)
$\boldsymbol{A}$	transition matrix of a SMPMM	subsec. 6.3.3 (p. 109)
$\boldsymbol{c}^\top$	selection vector for a MM	subsec. 4.4.1 (p. 50)
$\boldsymbol{c}_{PFD}^\top$	PFD selection vector for a SMPMM	def. 6.10 (p. 109)
$\boldsymbol{c}_{PFS}^\top$	PFS selection vector for a SMPMM	def. 6.10 (p. 109)
δ	delay specification for IEC 61508 ed. 2 petri net transitions	subsec. 4.5.3 (p. 62)
Δt	discrete time step size	subsec. 4.4.1 (p. 50)
dv_{SIS}	output vertex of a DASMLG	def. 7.2 (p. 124)
f_D	demand frequency	sec. 3.1 (p. 29)
$\boldsymbol{iv}$	vector of input vertices for sequences	def. 5.13 (p. 85)
$\boldsymbol{iv}$	vector of input vertices for voters	def. 5.14 (p. 85)
$\boldsymbol{iv}$	vector of input vertices for combinatorial sequences	def. 7.4 (p. 125)
k	discrete time step variable	
k_{mt}	mission time time steps for a SIS described as SMPMM	def. 6.1 (p. 106)
k_t	point in time for a phase transition in a SMPMM	def. 6.17 (p. 118)
k_{pt}	discrete time step at which a proof test is conducted for a SIS modeled as MPMM	def. 4.1 (p. 56)
λ	failure rate	
$\lambda_{(\cdot)}$	specific failure rate, $(\cdot)$ = e.g.ddx, du, ...	def. 5.11 (p. 83)
$\boldsymbol{M}_t$	phase transition matrix for a phase transition in an SMPMM	def. 6.17 (p. 118)
μ	repair rate	
$\mu_{(\cdot)}$	specific repair rate, $(\cdot)$ = e.g.ddr, dr, ...	def. 5.11 (p. 83)
N_{SIS}	number of components in a SIS	def. 6.7 (p. 108)
$\boldsymbol{p}(k)$	probability distribution of a MM over its state space at discrete time step k	subsec. 4.4.1 (p. 50)
$\boldsymbol{p}_0$	initial probability distribution for a MM	subsec. 4.4.1 (p. 50)
$\boldsymbol{p}_0$	initial probability distribution for a SMPMM	def. 6.9 (p. 108)
PFD_{avg}	average probability of failure on demand	sec. 3.1 (p. 29)
PFD_{avg}	average probability of failure on demand in SMPMMs	eq. 6.5 (p. 107)
PFD_{avg}	average probability of failure on demand in DASMLSISs	eq. 7.9 (p. 127)

Variable	Explanation	Reference
PFS_{avg}	average probability of fail-safe	sec. 3.2 (p. 31)
PFS_{avg}	average probability of fail-safe in SMPMMs	eq. 6.5 (p. 107)
PFS_{avg}	average probability of fail-safe in DASML-SISs	eq. 7.9 (p. 127)
Q	size of a MM's state space	def. 6.6 (p. 107)
QT	queuing time i.e.the time from a failure occurrence to the upcoming proof test	subsec. 3.5.2 (p. 36)
$\boldsymbol{R}$	probability shifting matrix for a MPMM	subsec. 4.4.5 (p. 55)
$\boldsymbol{R}_{mt}$	component reset matrix for a SMPMM describing a mission time replacement	def. 6.16 (p. 117)
$\boldsymbol{R}_{pt}$	component reset matrix for a SMPMM describing a proof test	def. 6.16 (p. 117)
$\boldsymbol{R}_{ppt}$	component reset matrix for a SMPMM describing a partial proof test	def. 6.16 (p. 117)
R_P	process risk	sec. 3.1 (p. 29)
R'_P	process risk disregarding alternative protection measures	sec. 3.1 (p. 29)
R_R	residual risk	sec. 3.1 (p. 29)
$\tilde{\boldsymbol{\sigma}}$	component state configuration vector	def. 5.21 (p. 93)
$\boldsymbol{s}_i$	state space vector i of a markov model	
$\boldsymbol{sdc}$	shutdown combination vector	def. 5.16 (p. 87)
$\tilde{t}$	arbitrary point in time where component state functions $\sigma_i(t = \tilde{t})$ have predefined values	def. 5.21 (p. 93)
Θ_{In}	input terminal for RBDs	subsec. 4.2.1 (p. 42)
Θ_{Out}	output terminal for RBDs	subsec. 4.2.1 (p. 42)
T_{obs}	observation period	
T_I	notation for a constant proof test interval according to IEC 61508	
TTF	random variable time-to-failure	sec. 3.3 (p. 32)
TTR	random variable time-to-restoration	sec. 3.4 (p. 33)
v_{SIS}	output vertex of an ASMLG	def. 5.10 (p. 82)
$\boldsymbol{vs}$	vector of voting schemes	def. 5.14 (p. 85)

Functions

Function	Explanation	Reference
Com	computation time function	sec. 7.4 (p. 134)
Diff	vector difference set function	def. 6.11 (p. 110)
DDC	generalized *dd* failure carrier function	def. 5.17 (p. 89)
DDC_{COM}	*dd* failure carrier function for components	def. 5.18 (p. 90)
DDC_{SEQ}	*dd* failure carrier function for sequences	def. 5.19 (p. 90)
DDC_{VOT}	*dd* failure carrier function for voters	def. 5.20 (p. 91)
Dec	decomposition function	def. 7.10 (p. 128)
$\mathrm{DT}(t)$	downtime i.e.total time spent in unavailability state	sec. 3.5 (p. 34)
EP	edge path function	def. 5.8 (p. 82)
EPD	edge path destination function	def. 5.9 (p. 82)
Inh	inhibition set function	def. 6.12 (p. 110)
ϕ	phase transition function	def. 6.2 (p. 106)
$\mathrm{F}(t)$	point unreliability; CDF of a lifetime distribution	sec. 3.3 (p. 32)
$\overline{\mathrm{F}}(T)$	unreliability averaged over T	
$\mathrm{M}(t)$	CDF of a restoration time distribution	sec. 3.3 (p. 32)
MAINT	supplementary function for DEC	def. 7.10 (p. 128)
Mem	memory consumption function	sec. 7.4 (p. 134)
MG	maintenance group function	def. 5.24 (p. 98)
PATH	supplementary function for DEC	def. 7.10 (p. 128)
PFD	probability of failure on demand function for DASMLGs	def. 7.5 (p. 126)
$\mathrm{PFD}(k)$	probability of failure on demand over discrete time i.e.safety related unavailability	def. 6.4 (p. 107)
$\mathrm{PFD}(k)$	probability of failure on demand over discrete time for DASMLSISs	def. 7.9 (p. 127)
$\mathrm{PFD}(t)$	probability of failure on demand over time i.e.safety related unavailability	
PFD_{CSEQ}	probability of failure on demand function for combinatorial sequences in DASMLGs	def. 7.7 (p. 126)
PFD_{SUB}	probability of failure on demand function for subsystems in DASMLGs	def. 7.7 (p. 126)
PFDC	generalized PFD contribution function	def. 5.17 (p. 89)
PFDC_{COM}	PFD contribution function for components	def. 5.18 (p. 90)
PFDC_{SEQ}	PFD contribution function for sequences	def. 5.19 (p. 90)
PFDC_{VOT}	PFD contribution function for voters	def. 5.20 (p. 91)
PFS	probability of fail-safe function for DASMLGs	def. 7.5 (p. 126)
$\mathrm{PFS}(k)$	probability of fail-safe over discrete time	def. 6.4 (p. 107)

Function	Explanation	Reference
PFS(k)	probability of fail-safe over discrete time for DASMLSISs	def. 7.9 (p. 127)
PFS(t)	probability of fail-safe over time	
PFS_{CSEQ}	probability of fail-safe function for combinatorial sequences in DASMLGs	def. 7.7 (p. 126)
PFS_{SUB}	probability of fail-safe function for subsystems in DASMLGs	def. 7.7 (p. 126)
PFSC	generalized PFS contribution function	def. 5.17 (p. 89)
PFSC_{COM}	PFS contribution function for components	def. 5.18 (p. 90)
PFSC_{SEQ}	PFS contribution function for sequences	def. 5.19 (p. 90)
PFSC_{VOT}	PFS contribution function for voters	def. 5.20 (p. 91)
σ	component state function	def. 5.12 (p. 84)
U(t)	point unavailability for an RFR process	sec. 3.5 (p. 34)
$\overline{\text{U}}(T)$	unavailability averaged over T	
UT(t)	uptime i.e.total time spent in availability state	sec. 3.5 (p. 34)
$v(t)$	repair intensity	eq. 3.9 (p. 35)
$w(t)$	unconditional failure intensity	eq. 3.9 (p. 35)

Sets

Set	Explanation	Reference
$\mathcal{AS}$	assertions for IEC 61508 ed. 2 petri net transitions	subsec. 4.5.3 (p. 62)
$\mathcal{B}$	blocks in RBDs	subsec. 4.2.1 (p. 42)
$\mathcal{COM}$	component vertices in an ASMLG	def. 5.3 (p. 78)
$\mathcal{DE}$	edges in a DASMLG	def. 7.1 (p. 123)
$\mathcal{DV}$	vertices in a DASMLG	def. 7.1 (p. 123)
$\mathcal{CS}$	component states	def. 5.1 (p. 72)
$\mathcal{CSEQ}$	combinatorial sequence vertices in a DASMLG	def. 7.1 (p. 123)
$\mathcal{E}$	edges in RBDs	subsec. 4.2.1 (p. 42)
$\mathcal{E}$	edges in FTs	subsec. 4.3.1 (p. 45)
$\mathcal{E}$	edges in ASMLGs	sec. 5.3 (p. 77)
$\mathcal{EP}$	edge paths of an ASMLG	def. 5.7 (p. 81)
$\mathcal{FT}$	failure types	def. 5.2 (p. 73)
$\mathcal{MG}$	maintenance group	def. 5.23 (p. 98)
$\mathcal{MGS}$	maintenance groups	def. 5.22 (p. 98)
$\mathcal{MT}$	points in time a component replacement is conducted due to mission time expiration	def. 5.11 (p. 83)
$\mathcal{P}$	primary failure events in FTs	subsec. 4.3.1 (p. 45)
$\mathcal{R}$	logical relations in FTs	subsec. 4.3.1 (p. 45)
$\mathcal{PR}$	predicates for IEC 61508 ed. 2 petri net transitions	subsec. 4.5.3 (p. 62)
$\mathcal{PT}$	points in time a proof test is conducted	def. 5.11 (p. 83)
$\mathcal{PPT}$	points in time a partial proof test is conducted	def. 5.11 (p. 83)
$\mathcal{REM}$	removable states due to optimization	def. 7.12 (p. 133)
$\mathcal{S}$	state space of a MM	subsec. 4.4.1 (p. 50)
$\mathcal{S}$	state space of an SMPMM	def. 6.8 (p. 108)
$\mathcal{SEQ}$	sequence vertices in an ASMLG	def. 5.3 (p. 78)
$\mathcal{SUB}$	subsystem vertices in a DASMLG	def. 7.1 (p. 123)
$\mathcal{T}$	phase transitions	def. 6.17 (p. 118)
$\mathcal{V}$	vertices in an ASMLG	def. 5.3 (p. 78)
$\mathcal{VOT}$	voter vertices in an ASMLG	def. 5.3 (p. 78)
$\mathcal{VS}$	voting scheme	def. 5.15 (p. 86)

Literaturverzeichnis

[BB95] N. Balakrishnan and A. P. Basu, editors. *The Exponential Distribution: Theory, Methods and Applications*. OPA (Ovewrseas Publishers Association), 1995.

[BL97] J. V. Bukowski and A. Lele. The case for architecture-specific common cause failure rates and how they affect system performance. volume 83, pages 153–158, Philadelphia, PA, January 1997. RAMS - Annual Reliability and Maintainability Symposium.

[BMS10] M. Blum, T. Mattes, and F. Schiller. Pfh-calculation for complex safety functions by means of generated markov models. volume 7, pages 49–52, Kassel, June 2010. INSS - International Conference on Networked Sensing Systems.

[Bö04] J. Börcsök. *Elektronische Sicherheitssysteme - Hardwarekonzepte, Modelle und Berechnung*. Hüthig GmbH & Co. KG, Heidelberg, 1st edition, 2004.

[Bö06] J. Börcsök. *Funktionale Sicherheit - Grundzüge sicherheitstechnischer Systeme*. Hüthig GmbH & Co. KG, Heidelberg, 1st edition, 2006.

[BSH08] J. Börcsök, B. Schrörs, and P. Holub. Reduzierung der ausfallwahrscheinlichkeit und verlängerung des proof-test-intervalls durch einsatz von partial-stroke-tests am beispiel von stellgeräten. *atp - Automatisierungstechnische Praxis*, 50(11):48–56, November 2008.

[Buk01] J. V. Bukowski. Modeling and analyzing the effects of periodic inspection on the performance of safety-critical systems. *IEEE Transactions on Reliability*, 50(3):321–329, September 2001.

[Buk05] J. V. Bukowski. A comparison of techniques for computing pfd average. volume 91, pages 590–595, Alexandria, VA, January 2005. RAMS - Annual Reliability and Maintainability Symposium.

[Buk06] J. V. Bukowski. Using markov models to compute probability of failed dangerous when repair times are not exponentially distributed. volume 92, pages 273–277, Washington DC, January 2006. RAMS - Annual Reliability and Maintainability Symposium.

[But86] R. W. Butler. An abstract language for specifying markov reliability models. *IEEE Transactions on Reliability*, 35(5):595–601, December 1986.

[BvHS09] J. Braband, R. vom Hövel, and H. Schäbe. Probability of failure on demand – the why and the how. volume 28, Hamburg, September 2009. SAFECOMP.

[CL08] C. G. Cassandras and S. Lafortune. *Introduction to Discrete Event Systems*. Springer Science+Business Media LLC, New York, NY, 2nd edition, 2008.

[DIN02] DIN EN 61508. Funktionale Sicherheit sicherheitsbezogener elektrischer / elektronischer / programmierbarer elektronischer Systeme, November 2002.

[DIN04] DIN IEC 61165. Anwendung des Markoff-Verfahrens, January 2004.

[DIN05] DIN EN 61511. Funktionale Sicherheit - Sicherheitstechnische Systeme für die Prozessindustrie, May 2005.

[DIN07] DIN EN ISO 13849-1. Sicherheit von Maschinen - Sicherheitsbezogene Teile von Steuerungen - Teil 1: Allgemeine Gestaltungsleitsätze, July 2007.

[DLN08] D. Düpont, L. Litz, and P. Netter. Lokalisierung und Analyse von Fehlerquellen beim numerischen SIL-Nachweis. *atp - Automatisierungstechnische Praxis*, 50(2):62–67, February 2008.

[Dü10] D. Düpont. *Merging Bottom-up and Top-down Availability for realistic Analysis of Safety-related Loops*. PhD thesis, University of Kaiserslautern, April 2010.

[Erm75] S. M. Ermakow. *Die Monte-Carlo-Methode und verwandte Fragen*. Oldenbourg Verlag, München, 1st edition, 1975.

[exi] exida.com LLC. exida - Functional Safety, Security & Reliability. URL: http://www.exida.com, last checked: 2010-07-25.

[FF10] F. Felgner and G. Frey. Effiziente modellierung zeitkontinuierlicher markovmodelle in modelica. volume 11, pages 245–254, Magdeburg, May 2010. Fachtagung Entwurf komplexer Automatisierungssysteme (EKA 2010).

[GC05] W. M. Goble and H. Cheddie. *Safety Instrumented Systems Verification*. ISA - The Instrumentation, Systems and Automation Society, Research Triangle Park, NC, 1st edition, 2005.

[GLS08a] T. Gabriel, L. Litz, and B. Schrörs. A formal approach to derive configurable markov models for arbitrarily structured safety loops. volume 9, Hong Kong, May 2008. PSAM - International Conference on Probabilistic Safety Assessment and Management.

[GLS08b] T. Gabriel, L. Litz, and B. Schrörs. Systematic Approach for the SIL-Proof of non-standard Safety-Loops. volume 5, Weimar, June 2008. Proceedings of the 5th Petroleum and Chemical Industry Conference Europe - Electrical and Instrumentation Applications.

[Gob00] W. M. Goble. *The Use and Development of Quantitative Reliability and Safety Analysis in New Product Design.* exida.com LLC, Sellersville, PA, 1st edition, 2000.

[Gul03] W. G. Gulland. Repairable redundant systems and the markov fallacy. Technical report, 4-Sight Consulting, 2003.

[GY08] H. Guo and X. Yang. Automatic creation of markov models for reliability assessment of safety instrumented systems. volume 93, pages 829–837, June 2008.

[Har09] J. Hartung. *Statistik - Lehr- und Handbuch der angewandten Statistik.* Oldenbourg Verlag, München, 15th edition, 2009.

[Hil07a] A. Hildebrandt. Berechnung der 'probability of failure on demand' (pfd) einer heterogenen 1-aus-2-struktur in anlehnung an die en 61508. *atp - Automatisierungstechnische Praxis*, 49(10):73–80, October 2007.

[Hil07b] A. Hildebrandt. Calculating the 'probability of failure on demand' (pfd) of complex structures by means of markov models. volume 4, Paris, June 2007. PCIC Europe - Petroleum & Chemical Industry Committee.

[HMT06] P. Hokstad, A. Maria, and P. Tomis. Estimation of common cause factors from systems with different numbers of channels. *IEEE Transactions on Reliability*, 55(1):18–25, March 2006.

[HR05] M. J. M. Houtermans and J. L. Rouvroye. The influence of design parameters on the probability of fail-safe (pfs) performance of safety instrumented systems (sis). Technical report, Safety Users Group, September 2005.

[HS99] M. J. Hassett and D. G. Stewart. *Probability For Risk Management.* ACTEX Publications, Inc., 1999.

[HT96] W. Hengartner and R. Theodorescu. *Einführung in die Monte-Carlo-Methode.* Carl Hanser Verlag, München, 1st edition, 1996.

[IDRS08] F. Innal, Y. Dutuit, A. Rauzy, and J.-P. Signoret. New insight into pfdavg and pfh. Technical report, Safety Users Group, July 2008.

[IEC09] IEC 61508-1 ed. 2. Functional safety of electrical/electronic/programmable electronic safety-related systems - Part 1: General requirements, 2009.

[Inn08] F. Innal. *Contribution to modelling safety instrumented systems and to assessing their performance - Critical analysis of IEC 61508 standard.* PhD thesis, University of Bordeaux, July 2008.

[JB88] S. C. Johnson and R. W. Butler. Automated generation of reliability models. volume 83, pages 17–22, Los Angeles, CA, January 1988. RAMS - Annual Reliability and Maintainability Symposium.

[KH96] H. Kumamoto and E. J. Henley. *Probabilistic Risk Assessment and Management for Engineers and Scientists.* IEEE Inc., New York, NY, 2nd edition, 1996.

[Lit98] L. Litz. Grundlagen der sicherheitsgerichteten automatisierungstechnik. *at - Automatisierungstechnik*, 46(2):56–68, February 1998.

[Lit01] L. Litz. *Wahrscheinlichkeitstheorie für Ingenieure - Grundlagen, Anwendungen, Übungen.* Hüthig GmbH & Co. KG, Heidelberg, 1st edition, 2001.

[Lit05] L. Litz. *Grundlagen der Automatisierungstechnik.* Oldenbourg Verlag, München, 1st edition, 2005.

[LR08a] M. A. Lundteigen and M. Rausand. The effect of partial stroke testing on the reliability of safety valves. *Journal of Loss Prevention in the Process Industries*, 21(6):579–588, November 2008.

[LR08b] M. A. Lundteigen and M. Rausand. Spurious activation of safety instrumented systems in the oil and gas industry: Basic concepts and formulas. volume 93, pages 1208–1217, August 2008.

[May09] T. Mayer. Generische Markov-Modell-basierte PFD-Berechnung für PLT-Schutzeinrichtungen mit dynamischem Voting unter Reparatureinfluss. Master's thesis, University of Kaiserslautern, 2009.

[MBC84] M. Ajmone Marsan, G. Balbo, and G. Conte. A class of generalized stochastic petri nets for the performance analysis of multiprocessor systems. *ACM Transactions on Computer Systems*, 2(1):93–122, May 1984.

[ML11] K. Machleidt and L. Litz. An optimization approach for safety instrumented system design. volume 97, Lake Buena Vista, FL, January 2011. RAMS - Annual Reliability and Maintainability Symposium. To be published.

[MS06] R. McCrea-Steele. Partial stroke testing - the good, the bad and the ugly. volume 7, Köln, May 2006. International Symposium on Programmable Electronic Systems in Safety Related Applications.

[OA10] L. F. Oliveira and R. N. Abramovitch. Extension of isa tr 84.00.02 pfd equations to koon architectures. volume 95, pages 707–715, July 2010.

[Pet62] C. A. Petri. *Kommunikation mit Automaten.* PhD thesis, University of Darmstadt, June 1962.

[PP98] J. Pukite and P. Pukite. *Modeling for Reliability Analysis.* IEEE Press Series on Engineering of Complex Computer Systems. IEEE Inc., New York, NY, 1st edition, 1998.

[PR99] G. Point and A. Rauzy. Altarica: Constraint automata as a description language. *Journal Européen des Systèmes Automatisés*, 33(8–9):1033–1052, March 1999.

[Rau04] A. Rauzy. An experimental study on iterative methods to compute transient solutions of large markov models. volume 86, pages 105–115, October 2004.

[RH04] M. Rausand and A. Hoyland. *System Reliability Theory - Models, Statistical Methods, and Applications.* Wiley Series in Probability and Statistics. John Wiley & Sons, Inc., Hoboken, NJ, 2nd edition, 2004.

[Rom05] J. L. Romeu. Availability. volume 24, pages 4–10, spring 2005.

[Rou04] J. L. Rouvroye. *Enhanced Markov Analysis as a method to assess safety in the process industry.* PhD thesis, Technische Universiteit Eindhoven, 2004.

[RvdB02] J. L. Rouvroye and E. G. van den Bliek. Comparing safety analysis techniques. *Reliability Engineering and System Safety*, 75(3):289–294, March 2002.

[Saf] U.S. Chemical Safety and Hazard Investigation Board. URL: http://www.csb.gov, last checked: 2010-07-22.

[Sch76] W. G. Schneeweiss. On the mean duration of hidden faults in periodically checked systems. *IEEE Transactions on Reliability*, 25(5):346–348, December 1976.

[Sch80] W. G. Schneeweiss. *Zuverlässigkeits-Systemtheorie - Methoden zur Beurteilung der Zuverlässigkeit technischer Systeme.* Datakontext-Verlag GmbH, Köln-Bonn, 1st edition, 1980.

[Sie95] M. Siegle. *Beschreibung und Analyse von Markovmodellen mit großem Zustandsraum.* PhD thesis, Universität Erlangen - Nürnberg, 1995.

[SN 07] SN 31920. Standard B10-Werte bei kontinuierlicher Anforderungsrate und Ausfallraten bei niedriger Anforderungsrate elektromechanischer Komponenten, June 2007.

[TEMT09a] A. C. Torres-Echeverria, S. Martorell, and H. A. Thompson. Design optimization of a safety-instrumented system based on rams+c addressing iec61508 requirements and diverse redundancy. volume 94, pages 162–179, February 2009.

[TEMT09b] A. C. Torres-Echeverria, S. Martorell, and H. A. Thompson. Modelling and optimization of proof testing policies for safety instrumented systems. volume 94, pages 838–854, April 2009.

[VPH05] W. Velten-Philipp and M. J. M. Houtermans. The effect of diagnostic and periodic testing on the reliability of safety systems. volume 1, Beachwood, OH, June 2005. International Safety Symposium USA.

[Wag03] C. Wagenknecht. *Algorithmen und Komplexität*. Carl Hanser Verlag, 1st edition, 2003.

[Wei09] C. Weinmann. Erstellung eines mathematischen Modells zur Bestimmung sicherheitstechnischer Kenngrößen für automatisierte Prozessanalysentechnik. Master's thesis, University of Kaiserslautern, 2009.

[Win08] H. Winter. *Prozessleittechnik in Chemieanlagen*. Europa-Lehrmittel, Nourney, Vollmer GmbH & Co. KG, Haan-Gruiten, 3rd edition, 2008.